An introduction to the physical theory

RESPONSE AND STABILITY

A.B.PIPPARD FRS

Emeritus professor of physics, University of Cambridge

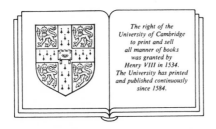

*The right of the
University of Cambridge
to print and sell
all manner of books
was granted by
Henry VIII in 1534.
The University has printed
and published continuously
since 1584.*

CAMBRIDGE UNIVERSITY PRESS

Cambridge

London   New York   New Rochelle

Melbourne   Sydney

Published by the Press Syndicate of the University of Cambridge
The Pitt Building, Trumpington Street, Cambridge CB2 1RP
32 East 57th Street, New York, NY 10022, USA
10 Stamford Road, Oakleigh, Melbourne 3166, Australia

First published 1985
Reprinted with corrections 1986

Printed in Great Britain at the University Press, Cambridge

Library of Congress catalogue card number: 85-1304

British Library cataloguing in publication data

Pippard, A.B.
Response & stability: an introduction to physical theory.
1. Equilibrium   2. Mechanics
I. Title
531'.12   QA831

ISBN 0 521 26673 4 hard covers
ISBN 0 521 31994 3 paperback

MP

CONTENTS

[*for the meaning of square brackets, see p. xi*]

PREFACE

This book is a much expanded, though not very much more advanced, version of a short course of lectures I gave for some years to third-year physics students in Cambridge, in an attempt to broaden their outlook. Though well instructed in the central themes of physics – quantum mechanics, thermodynamics, solid state, etc. – they had only incidentally met examples of instabilities, and of these the most atypical, phase changes, had been most emphasized. As for serious discussion of non-linear response, this was almost entirely neglected. I suspect the same is generally true elsewhere. So I tried to indicate, with demonstrations, the unifying principles, such as they are; whether I succeeded or not, I certainly convinced myself that there are many things in this field that the practising physicist ought to know, especially the physicist who intends to work outside the research laboratories where narrow specialization is virtually essential to success. Engineers surely do not need to be told much of what I have to say, but I hope they will derive pleasure from reading what they already know, told differently. Many of the illustrative examples concern the engineer as much as the physicist even if, inevitably, they are dished up with a physical flavour. Only in the last two chapters are the topics slanted heavily towards the student of physics. Here, as elsewhere, I have adopted the prolix style which is the only way I can learn difficult concepts. A more economical treatment could compress most of the mathematical theory into a few pages, but it would take an exceptional student to understand the applications of the simple algebra without the examples. In any case, I find the mathematics rather dull and the physical and engineering examples interesting, and I write for those who share my tastes.

My best thanks are due to Basil Humphreys for his skill in making the sphere of fig. 2.11, to Norman Bett for help with electronics, especially in setting up the circuit of fig. 4.28 and the measuring equipment to go with it, to Owen Saxton for correcting my treatment of Ex. 5.11, to Ken Riley for his critical comments on the first two chapters, and to Andrew Phillips for helpful discussion of chapter 7. The photographs were prepared by Keith Papworth and Eleanor Pippard, and the manuscript was deciphered and typed by Susan Marriott and Margaret Brown. To these also I convey my gratitude.

9.10.84   *A. B. Pippard*

INTRODUCTORY REMARKS

which are rather important

The idea of continuity is deeply embedded in the minds of scientists, especially physicists. This is as it should be when the only successful way of describing processes exactly (and physics *is* successful) is dominated by differential equations – if functions are not continuous, they have no differential coefficients. Continuity is not, however, the universal rule and from time to time sudden discontinuities make their appearance. Water on the stove suddenly begins to boil (especially if the dissolved air has been removed), things fall over at a touch, the air is suddenly rent by a stroke of lightning. While textbooks of physics normally stress continuous processes, if only because they are easier to understand, and must precede the study of discontinuity, this book is far more concerned with the moments of sudden change. It addresses questions like these:

> What are the criteria by which, without doing an experiment, one can decide whether a given system is in stable equilibrium?
>
> If a system in stable equilibrium is disturbed, by what path does it return to its original state?
>
> When the parameters of a system are slowly changed, in what circumstances does the configuration change sharply, and what can one say about subsequent events?
>
> When a system in equilibrium is disturbed by a periodic force, will it respond at the same frequency, some other frequency, or aperiodically?

These are questions which are apt to arise in all branches of physics and engineering, and not only in scientific circles. The same principles are called upon when one seeks to understand the ability of a seal to balance a ball on its nose and a human being to ride a bicycle, the gravitational collapse of a star, or the breaking of the Tacoma bridge – these are all problems of stability. Similarly, rather recently developed ideas about chaotic response find application in the growth of insect populations and in weather prediction.

None of these applications will be discussed, but many others will be. In fact, much of the book is taken up with specific examples illustrating different phenomena in a wide variety of contexts, yet having certain aspects of their behaviour in common. Too much must not be expected, however; there is a limit to the extent to which general procedures can answer all the questions, and one must frequently devise special methods to tackle each problem as it arises. In order to give practice in problem-solving I have interspersed exercises, some of which are simply problems of the conventional sort, to test your appreciation of the principles. In addition I have left a considerable number of calculations incomplete, or in skeleton form, indicating the processes by which they can be completed. It is my hope that by working through these you will develop your competence in expressing physical conceptions in mathematical form, and in following the logic through to the end. Ideally, anyone learning physics should have unlimited access to an experienced teacher, from whom he can discover his mistakes and get practice in the techniques that only practice can impart. I have kept in mind, while writing, the imperfect nature of things, and have tried to give hints and encouragement that I fear can only be a faint shadow of the living ideal. But if you are worried that I am setting a herculean task of problem-solving just to understand the drift of the arguments, be reassured – whenever the answer to an exercise is needed for the subsequent discussion, the answer is given. It is therefore possible to skim through, though at the risk of imperfect mastery. Further, you need not be discouraged if you find some of the exercises or theoretical discussion difficult – just leave out the difficult parts and carry on; the chances are that you will be able to pick up the thread again quite easily. The cross-references in the margin should help if you want to skip.

From time to time I describe, or give reference to, experiments which do not require much, or expensive, apparatus; and I quote results I have obtained myself, warts and all. They are all worth doing yourself but, again, if you can't manage it, you will not lose the thrust of the argument. What you lose by neglecting experiment is the sense of immediacy and, a more subtle matter, the realization that a piece of equipment is not simply a set of equations, but may have its own ideas about how to behave which do not quite fit into the mathematical model. In physics, perhaps especially in the branches touched on here, mathematics is apt to be at a loss unless there are experiments to point the way. This is not to imply that a first-class mathematician might not sometimes find the answers without the aid of experiment, but he would take longer than a mediocre mathematician who realized the benefit to be gained from a few practical trials.

In a few places, on pp. 36–38, 127–128, 169–170, 216–219, you will find passages enclosed in square brackets. Here I have allowed myself, among other things, the luxury of speculating on ways in which the ideas might be applied outside the realm of physics. You are not, of course, expected to treat these as anything more than personal views, but I hope they will stimulate you to think about how physical concepts may provide models (I mean analogies, not mathematical simulations) of processes far beyond the capability of exact analysis, but not necessarily beyond rational discussion.

1

The harmonic oscillator

You will, of course, already know something of simple harmonic motion, but a little revision will do no harm, and how better than by solving a few problems? But first let us remind ourselves that the characteristic differential equation for the simplest possible form of oscillator is

$$\ddot{x} + \omega_0^2 x = 0. \tag{1.1}$$

A mass m hanging on a spring obeys this equation. If (a) in fig. 1.1 represents the mass in equilibrium, when it is pulled downwards a distance y it will experience a restoring force, in the upward direction, proportional to y, μy say. Then the equation of motion is

$$m\ddot{y} = -\mu y,$$

having the same form as (1.1) if $(\mu/m)^{\frac{1}{2}} = \omega_0$, the angular frequency of vibration. Similarly if q is the charge on the capacitor in fig. 1.1(c), and i the current,

$$\dot{q} = i.$$

Also the potential drop in a clockwise direction is q/C across C and $L\, di/dt$, i.e. $L\ddot{q}$ across L. Since there must be no change of potential going round the circuit,

$$L\ddot{q} + q/C = 0, \tag{1.2}$$

giving, by comparison with (1.1), a frequency $\omega_0 = (LC)^{-\frac{1}{2}}$.

The general solution of (1.1) may be conveniently expressed in terms of complex numbers:

$$x = Ae^{i\omega_0 t} + Be^{-i\omega_0 t}, \text{ in which } A \text{ and } B \text{ are complex.}$$

If x is to represent a real physical quantity, A and B must be chosen so that x is real for all t, i.e. $x^* = x$. Thus

$$(A - B^*)e^{i\omega_0 t} + (A^* - B)e^{-i\omega_0 t} = 0,$$

which can be true at all times only if $B^* = A$. The most general real solution of (1.1) is therefore $Ae^{i\omega_0 t}$ + complex conjugate, or $2\,\mathrm{Re}\,(Ae^{i\omega_0 t})$. Obviously the factor 2 can be incorporated in A.

The complex, time-dependent quantities $Ae^{i\omega_0 t}$ and $A*e^{-i\omega_0 t}$ are represented on the Argand diagram as two vectors of length $|A|$ rotating anticlockwise and clockwise respectively with angular frequency ω_0. They are shown in fig. 1.2 at the instant $t=0$, and their sum lies on the real axis,

Fig. 1.1. A mass m, hanging from a spring, is displaced from equilibrium (a) downwards a distance y (b) and released. A loss-free LC circuit (c) carries a charge $\pm q$ on the capacitor plates, and a current i.

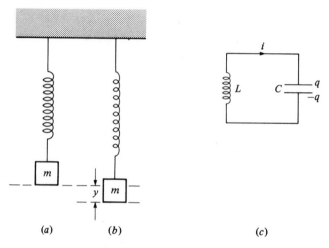

(a)   (b)   (c)

Fig. 1.2. The rotating vector $Ae^{i\omega t}$ and its complex conjugate $A*e^{-i\omega t}$.

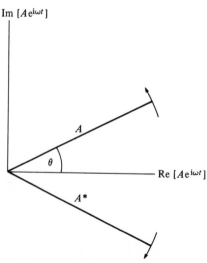

being twice the projection of either onto that axis. When A is real the vibration is at its positive extreme at $t = 0$, while when A is positive imaginary the vibration is at its central position, passing through it from positive to negative with a velocity $-\omega_0|A|$. If we express A as $|A|e^{i\theta}$, the vibration of the co-ordinate x takes the form

$$x = 2 \, \mathrm{Re} \, (|A|e^{i(\omega_0 t + \theta)}) \quad \text{or} \quad 2 \, \mathrm{Re} \, (|A|e^{-i(\omega_0 t + \theta)}),$$
$$= 2|A| \cos(\omega_0 t + \theta),$$

θ being seen as a phase advance relative to the 'standard' vibration $\cos \omega_0 t$. It does not matter whether one uses $e^{i\omega_0 t}$ or $e^{-i\omega_0 t}$ as the complex expression of a real vibration, provided one is consistent during the course of a calculation. I shall use both, more or less indiscriminately, as the whim takes me.

Determining the vibration frequency

The values of ω_0 for the simple examples in fig. 1.1 were found by writing the differential equation for one co-ordinate. Another method which is often useful depends on the fact that for loss-free simple harmonic motion (but for no other), any variable vibrating with amplitude a passes through its equilibrium position with velocity $\omega_0 a$. At this instant the energy of the system, measured from the equilibrium state of rest, is entirely kinetic; on the other hand, a quarter of a cycle later the variable is stationary at its greatest extent and the energy is entirely potential. The frequency may be found by writing down the condition that the energy is the same in the two states. Thus to depress the mass in fig. 1.1 through a height a work $\frac{1}{2}\mu a^2$ must be done, and this is the potential energy of the mass and the total energy if it is at rest. On being released it vibrates with frequency ω_0 and passes through the equilibrium position with kinetic energy $\frac{1}{2}m\omega_0^2 a^2$. For these two energies to be equal, $\mu = m\omega_0^2$, giving $\omega_0 = (\mu/m)^{\frac{1}{2}}$ as before.

Exercise 1.1. Show in the same way that a simple pendulum of length l has $\omega_0 = (g/l)^{\frac{1}{2}}$. The potential energy here is gravitational.

Exercise 1.2. Use an analogous method to show that the circuit of fig. 1.1(c) has frequency $(LC)^{-\frac{1}{2}}$. Note that if q is the maximum charge on the capacitor, the current one-quarter cycle later is $\omega_0 q$.

The next three examples should be solved both by the energy method and by direct analysis of forces and accelerations.

Exercise 1.3. A diatomic molecule may be thought of, for some purposes, as two atoms of mass m_1 and m_2 held a certain distance apart in equilibrium by forces that behave, for small displacements, like a spring of spring constant μ. Show that the vibration frequency is $[\mu_0(1/m_1 + 1/m_2)]^{\frac{1}{2}}$.

Exercise 1.4. In fig. 1.3 the pulley A turning smoothly on a horizontal axle is light but has a heavy semicircular rim B, of mass M. A string passing round the pulley carries a mass m. Draw a graph showing how the frequency of small vertical oscillations of m depends on the value of m. Make sure you understand why there is a critical mass m_c at which the behaviour changes. What happens when $m > m_c$? Show that near the critical point $\omega_0 \propto (m_c - m)^{\frac{1}{4}}$.

Exercise 1.5. A light cylinder has a heavy mass attached within it as shown in fig. 1.4. What is the frequency of small amplitude rolling oscillations if there is no slipping? It is helpful to remember that the point of contact with the surface is the instantaneous centre of rotation. If the contact could be made frictionless, what would the frequency be?

Exercise 1.6. This is somewhat similar to Ex. 1.5, and it is more important to examine the critical behaviour than to calculate the frequency. The long plank shown in fig. 1.5 has its centroid at C

Fig. 1.3. The mass m displaces the pulley to a new equilibrium position at θ, about which it can execute small oscillations.

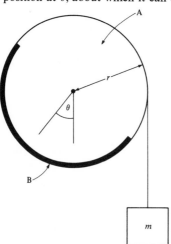

and rests on a cylinder. Show that the frequency of rocking motion goes to zero when R is reduced to equal H. If $R = H + x$, show that $\omega \propto x^{\frac{1}{2}}$ when x is small (cf. Ex. 1.4).

Lossy vibrators

If a resistor R is inserted in the circuit of fig. 1.1(c), an extra term $R\dot{q}$, appears on the left-hand side:

$$L\ddot{q} + R\dot{q} + q/C = 0,$$

a special case of the equation

$$\ddot{x} + 2\lambda\dot{x} + \omega_0^2 x = 0. \tag{1.3}$$

This is a linear differential equation, containing no powers of x, $\dot{x}$, $\ddot{x}$, etc., and no constant term, and the coefficients are constants. All such

Fig. 1.4. The dark circle represents a heavy mass firmly attached within a light cylinder. The whole executes rolling oscillations.

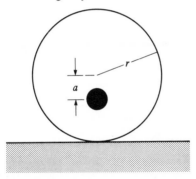

Fig. 1.5. A plank resting on a cylinder.

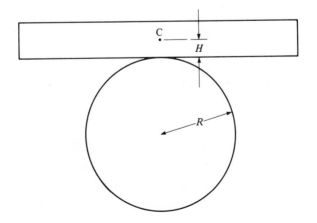

equations have exponential solutions. When $x = e^{pt}$ is substituted as a trial solution, $d^n x/dt^n = p^n x$ and an algebraic equation for p results. In this case

$$p^2 + 2\lambda p + \omega_0^2 = 0,$$

with two independent solutions,

$$p = -\lambda \pm (\lambda^2 - \omega_0^2)^{\frac{1}{2}}. \tag{1.4}$$

If $\lambda > \omega_0$ the solutions are non-oscillatory, but if $\omega_0 > \lambda$, the term $(\lambda^2 - \omega_0^2)^{\frac{1}{2}}$ is imaginary, meaning that the system oscillates with frequency $(\omega_0^2 - \lambda^2)^{\frac{1}{2}}$, but the oscillation decays as $e^{-\lambda t}$.

> *Exercise 1.7.* On an Argand diagram plot how p varies as λ changes. Although at present we are not concerned with negative values of λ, let it run between $\pm\infty$. There are several points to note, and to prove if you can from (1.3):
>
> (a) So long as p has an imaginary part, the two solutions for a given λ are complex conjugates. Hence show how to write down a real solution.
>
> (b) Show also that the values of p lie on a circle of radius ω_0.
>
> (c) But if p is real ($\lambda > \omega_0$), one solution lies inside the circle and one outside; $p_1 p_2 = \omega_0^2$.
>
> (d) If the oscillator is held at rest, distant x_0 from the centre, and let go, derive an expression for its subsequent motion. Taking $\lambda = \frac{1}{10}\omega_0$, sketch the solution.
>
> (e) Suppose in (d) the oscillator swings through zero but only rises to a peak of $x_0/100$ on the other side, what is now the value of λ? Find the point on the Argand diagram corresponding to this behaviour. Are you surprised how far it is from the point of critical damping, $\lambda = \omega_0$, above which there are no oscillations?
>
> (f) An overdamped system, with two real values of p, p_1 and p_2, is initially at rest when it is struck sharply to give it a velocity v_0. Find and sketch the solution that satisfies the initial conditions $x = 0, \dot{x} = v_0$. Note that it is the larger p that dominates at first, but the smaller p that determines the long-term behaviour.

According to (1.4) a damped oscillator has an amplitude that decays as $e^{-\lambda t}$:

$$x = x_0 e^{-\lambda t} e^{i\omega' t}, \quad \text{where} \quad \omega' = (\omega_0^2 - \lambda^2)^{\frac{1}{2}}.$$

The energy is proportional to the square of the amplitude, i.e. to $(x_0 e^{-\lambda t})^2$ and decays as $e^{-2\lambda t}$. The time constant τ_a during which the amplitude decays by a factor e is $1/\lambda$, twice as great as the time constant for the decay of energy, $\tau_e = 1/2\lambda$. The *quality factor*, Q, is defined as $\omega' \tau_e$; it is the number of radians of oscillation required for the energy to fall by a factor e.

If the damping is small, one may calculate it by estimating the dissipation directly. Thus for the *LCR* circuit one assumes a current $i = i_0 \cos \omega_0 t$, where $\omega_0 = (LC)^{-\frac{1}{2}}$. The instantaneous rate of dissipation is Ri^2, i.e. $Ri_0^2 \cos^2 \omega_0 t$, and, since the mean value of $\cos^2 \theta$ is $\frac{1}{2}$, the mean rate of dissipation is $\frac{1}{2}Ri_0^2$. Now the energy of the oscillation, $E = \frac{1}{2}Li_0^2$, so that $\dot{E} = -(R/L)E$. This equation has the solution $E = E_0 e^{-\mu t}$, where $\mu = R/L$. The time constant τ_e is consequently L/R, or $1/2\lambda$, as already found, since $R/L = 2\lambda$.

This device becomes useful when the damping is non-linear. It is perhaps not always appreciated that the model equation (1.3) is rarely realized in practice except for electrical circuits. When a pendulum swings in air it usually moves too fast for streamline motion, and when a turbulent wake accompanies the pendulum bob the frictional force varies with v according to a law lying somewhere between v and v^2. It is usually not possible to solve a differential equation like (1.3) in closed form if the middle term is a power or other non-linear function of x, and the calculation of dissipation is the most practicable way of finding how the oscillation decays. The method is reliable so long as the frictional term is not so strong as to distort the sinusoidal motion significantly.

Exercise 1.8. Consider a damped mechanical oscillator obeying the equation
$$m\ddot{x} + f(\dot{x}) + \mu x = 0,$$
in which $f(\dot{x})$ is:

(a) a constant retarding force (i.e. in the opposite direction to $\dot{x}$), such as is produced by sliding friction,

(b) a retarding force proportional to $\dot{x}^2$,

(c) a retarding force of the form $a|\dot{x}| + b\dot{x}^2$.

The curves in fig. 1.6 represent these three cases. You should be able to identify which belongs to each of the three friction laws without calculation, but it is worth calculating one or two cases anyway.

The constant frictional force (case (a)) when inserted as $c\dot{x}/|\dot{x}|$ instead of $2\lambda\dot{x}$ in (1.3) allows an explicit solution. See if you can find it, and compare the pattern of decay with that derived from the dissipation.

Driven harmonic oscillator

When a force proportional to $\cos \omega t$ is applied to an oscillator obeying (1.3), the equation of motion takes the form

$$\ddot{x} + 2\lambda\dot{x} + \omega_0^2 x = \mathrm{Re}\,(f e^{i\omega t});$$

alternatively expressed, the solution is the real part of the solution of

$$\ddot{x} + 2\lambda\dot{x} + \omega_0^2 x = f e^{i\omega t}. \tag{1.5}$$

The general solution can be expressed as a transient term which is $A e^{p_1 t} + B e^{p_2 t}$, as if there were no driving force, to which must be added any solution of (1.5). Such a solution is found by substituting $x = x_0 e^{i\omega t}$. Then

$$(-\omega^2 + 2i\omega\lambda + \omega_0^2)x_0 = f, \quad \text{and } x_0 \text{ is found.}$$

For the moment we shall ignore the transient terms and concentrate on the steady state response

$$x = f e^{i\omega t}/(\omega_0^2 - \omega^2 + 2i\omega\lambda). \tag{1.6}$$

Now it will be observed that the denominator vanishes when

$$\omega_0^2 - \omega^2 + 2i\omega\lambda = 0, \tag{1.7}$$

that is, when ω is one of the solutions, ω_1 or ω_2, which are the complex frequencies of free oscillation (cf. the solution of (1.3)). Hence the denominator may be factorized as $-(\omega - \omega_1)(\omega - \omega_2)$, and the steady state solution (1.6) takes the form

$$\begin{aligned} x &= -f e^{i\omega t}/(\omega - \omega_1)(\omega - \omega_2) \\ &= \frac{f e^{i\omega t}}{\omega_1 - \omega_2}\left[\frac{1}{\omega_1 - \omega} - \frac{1}{\omega_2 - \omega}\right]. \end{aligned} \tag{1.8}$$

In general, both terms in the square brackets must be retained. The frequencies, ω_1 and ω_2, of free oscillation have the same positive imaginary part, describing the decay of oscillation, and the real parts are equal and opposite, as illustrated on the Argand diagram of fig. 1.7(a). The two denominators are also shown, and it is clear that when $\mathrm{Im}\,[\omega_{1,2}]$ is small the contribution of $(\omega_1 - \omega)^{-1}$ is overwhelming in the vicinity of resonance.

Exercise 1.9. Show that if ω is real, $(\omega_1 - \omega)^{-1}$ lies on a circle of diameter $1/\mathrm{Im}\,[\omega_1]$, as in fig. 1.7($b$).

Fig. 1.6. Which of these tracings of damped oscillations corresponds to which of the three forms of frictional force in Ex. 1.8?

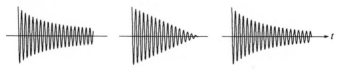

This illustrates the resonant response, both in amplitude and phase, the phase angle of $(\omega_1 - \omega)^{-1}$ being equal and opposite to that of $\omega_1 - \omega$. As ω runs from 0 to ∞, the response traverses the greater part of the circle and the phase of response changes through very nearly π.

The sharp resonant response of a low-loss system is called *Lorentzian*. If the second term in (1.8) is ignored, and (1.7) used to give the approximate solution (for small λ, $\omega_1 \approx \omega_0 + i\lambda$) the amplitude of response is seen to be

$$x \approx (-f/2\omega_0)/(\Delta\omega - i\lambda), \quad \text{where } \Delta\omega = \omega - \omega_0. \tag{1.9}$$

At low driving frequencies, $\Delta\omega < 0$ and $|\Delta\omega| \gg \lambda$; the response is in phase with the force, showing that the inertia of the vibrating object is unimportant, and the applied force serves to displace it against the restoring force. At high driving frequencies, $\Delta\omega$ is large and positive; the restoring force is of small importance and it is the inertia that must be overcome. At resonance, displacement and force are in phase quadrature, and the force is in phase with the velocity of the driven system. This allows the maximum rate of energy input to make up for the dissipation which of course is greatest at the resonant peak.

If we are concerned only with the magnitude of the response, and not with its phase, it is convenient to introduce the *intensity*, W, of response, defined as xx^*, the square of the amplitude. For Lorentzian response,

$$W = W_0/(\Delta\omega^2 + \lambda^2), \tag{1.10}$$

Fig. 1.7. (a) ω_1 and ω_2 on an Argand diagram. (b) $\omega_1 - \omega$ and $(\omega_1 - \omega)^{-1}$. As $\omega_1 - \omega$ traverses a horizontal line, its reciprocal traverses the circle.

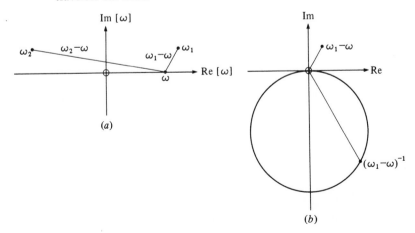

and

$$W_0 = f^2/4\omega_0^2.$$

W drops to one-half of its peak value when $\Delta\omega = \pm\lambda$. The width of resonance, measured at half the peak intensity, is therefore 2λ, or ω_0/Q.

> *Exercise 1.10.* A mass attached to a spring and subject to slight viscous damping is acted upon by an oscillating force. Show that when the frequency is tuned to give the maximum response, the response will be Q times the amplitude that would result from the same force acting on the mass alone, without the spring.

> *Exercise 1.11.* You will have observed that after the initial transients have died away the system always responds at the frequency of the driver. Show that this is true not only for a simple harmonic oscillator, but for any system where free motion is described by a linear differential equation with constant coefficients
>
> $$ax + b\dot{x} + c\ddot{x} + d\dddot{x} + \ldots = 0.$$

> We shall find later that this is not true in general for systems obeying non-linear equations.

2

Generalized linear systems and stability criteria

If you solved Ex. 1.7 correctly, you drew a diagram like fig. 2.1 to illustrate
the solutions $(x \propto e^{pt})$ of

$$\ddot{x} + 2\lambda \dot{x} + \omega_0^2 x = 0$$

as λ runs between $\pm \infty$. When $\lambda = 0$, x oscillates without damping (A and A'). As λ increases, Re (p) becomes negative, causing the oscillations to decay exponentially, as at B and B', until at C, the point of critical damping, oscillatory solutions are replaced by real exponential damping and typical solutions are D and D'. You will already have noted in Ex. 1.7(e) how heavily damped the oscillations are long before C is reached. From a physical point of view C represents no sudden change in behaviour, although mathematically it marks a change from one form of solution to another, as the trajectory turns sharply through a right angle.

Systems which return to equilibrium, after being disturbed, by a decaying oscillation are said to be in a state of *focal stability*, while those that return without oscillation are in *nodal stability*.

> *Exercise 2.1.* A certain overdamped system (i.e. one in a state of nodal stability) has p_1 and p_2 both negative, and $p_2 = 2p_1$. It is set in motion with $x = 1$ and with different velocities $\dot{x}$. Draw graphs of $x(t)$ for various $\dot{x}$, both positive and negative. For what range of values for $\dot{x}$ does x become negative?

Negative dissipation

It is possible to design simple electronic circuits, containing amplifiers, that behave as negative resistances. By the use of such devices λ may be caused to become negative. An example is shown in fig. 2.2(a). The gain of the operational amplifier is so large that for many purposes it may be assumed that A and B are always at the same potential. In an ideal amplifier no current flows in at A and B, while other leads (not shown) allow (within limits) whatever current is needed to come out at C; the flow of current is as shown.

> *Exercise 2.2* Show that the resistance between X and Y is $-R_3R_1/R_2$. This assumes the current is insufficient to saturate the amplifier.

> *Exercise 2.3.* If you are competent in electronics, you may think it is obvious that the feedback lead DB in fig. 2.2 must be connected to the negative input B. In this case you are probably accepting what you have been taught without thinking. In order to discuss

Fig. 2.1. Solutions of (1.3) over the whole range of λ, expressed as e^{pt} with p displayed on an Argand diagram. Positive λ on the left, negative on the right.

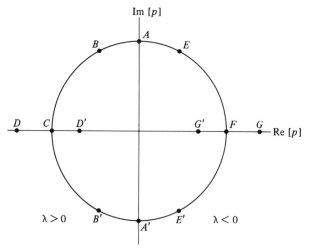

Fig. 2.2. (*a*) Circuit for providing negative resistance between X and Y. The additional notional inductance in the feedback lead of (*b*) enables the stability of the circuit to be analysed.

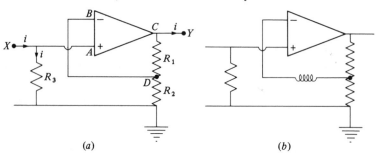

the question sensibly, it is necessary to make the circuit a little less idealized, so that the process shall take some time to be accomplished, and can be followed in detail. As in fig. 2.2(*b*), put a small inductor in the feedback loop, and in addition allow the amplifier to be a little less than perfect, i.e. let $V_C = \beta(V_A - V_B)$, where β is large but finite. You should now find that the equations of motion confer stability when D is connected to B, but instability when it is connected to A. Verify your result in the laboratory.

By connecting such a device in an *LCR* circuit, as in fig. 2.3, and adjusting R_1/R_2 or R_3, λ can be adjusted at will between positive and negative limits. This arrangement is well worth setting up. The only part that takes time is making the inductor L, since you should avoid using iron or ferrite cores to increase the inductance; these are apt to cause slightly non-linear behaviour and upset the experiment. The precise dimensions are of no consequence, and I have had good results by using a plastic bobbin, such as commercial copper wire is wound on, 5 cm long, 2 cm diameter, scramble-wound on a lathe with about 1500 turns of 30 SWG enamelled copper wire. Keep the inductor well away from iron, such as the cases of other components of the circuit.

> *Exercise 2.4.* If $L = 20$ mH, $C = 2$ μF, $R = 60$ Ω, $R_1 = 10^2$ Ω, $R_2 = 10^4$ Ω, what valuè of R_3 is needed for the circuit oscillation neither to grow nor decay? What change in R_3 is needed for the amplitude of oscillation to decay with a time constant of 1 s?

Fig. 2.3. The circuit of fig. 2.2 connected in series in an *LCR* circuit.

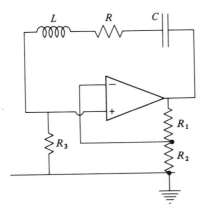

When the negative resistance is more than enough to annul R, the oscillation will grow exponentially (until saturation of the amplifier limits the amplitude). This corresponds to a point on the right semicircle in fig. 2.1, and the initial behaviour is termed *focal instability*. We are not ready yet to discuss the new stable state resulting from saturation.

72

> *Exercise 2.5.* Using the magnitudes of Ex. 2.4, calculate what R_0 must be to reach the critical point F, at which growing oscillations give way to real positive exponential growth (*nodal instability*).

> *Exercise 2.6.* Repeat Ex. 2.1 with p_1 and p_2 both positive.

Saddle-point instability

We have now discussed a harmonic vibrator with positive or negative damping. In both cases ω_0^2 was positive; a displaced particle suffered a restoring force, like a ball at the bottom of a curve. Suppose now ω_0^2 is allowed to be negative in (1.3), as would happen if the ball were resting at the top of a curve. Is there any form of viscous friction, positive or negative (treacle or anti-treacle) that will stabilize it at the top? Commonsense tells us, no; treacle will delay the fall, anti-treacle will accelerate it, but nothing will reverse the sign. The solution of (1.3) confirms this. If the solutions are written as e^{pt}, both values of p are always real, whatever the value of λ. And, since the product of the solutions is ω_0^2, i.e. negative, one p must be positive and the other negative. The positive p guarantees exponential growth of x at large t.

A system that can be adjusted so that ω_0^2 may be of either sign is the *Euler strut*.[1] A simple construction to study this important system is described in fig. 2.4. So long as the load is less than a certain critical mass m_c, the equilibrium position of the strut is vertical, but above m_c the vertical position is unstable, and there are two new stable positions, one either side of vertical. A critical point where one solution gives way to two is called a *bifurcation*, and we shall have a great deal more to say of these in due course.

131
102
133

Let us calculate the frequency of oscillation of the strut by determining the restoring force when it is displaced. We apply a horizontal force F, as in fig. 2.5, and find the resulting small displacement y_0. We take the load m as a point mass. The bending moment at P, with co-ordinates (x, y), is the

Fig. 2.4. Simple equipment to study the Euler strut. Some of the features visible in the photograph reveal that it was thrown together from what happened to be lying around in the workshop. There is little critical in the design except the strut itself which is made of a straight length of spring steel (Chesterman tape), $\frac{1}{2}''$ wide and 0.010″ thick. The free length of tape (10 cm in this case) can be adjusted to suit the load, which here is made of two pieces of brass $1'' \times \frac{3}{4}'' \times \frac{5}{16}''$. There are ball bearings under the two side arms and an adjustable screw at the back, to give a three-point steady support. The adjustable screw enables the strut to be made to vibrate symmetrically, even though it may not be perfectly straight. The extra loads, in the form of thin plastic strips, are shown at the side. They are slit to allow them to be slipped in place on top of the fixed brass load. When there are enough to bring the strut very near to the critical state, it is important to use very small amplitudes in measuring the period, since the quadratic region of potential is extremely limited.

couple that would have to be applied to hold the top part in place if the strut were cut at P. It is clearly $mg(y_0 - y) + F(l - x)$, where l is the length of the strut. This is in fact provided by the lower part of the strut, which exerts a couple $EI\, d^2y/dx^2$ when bent, if $y_0 \ll l$. Here E is Young's modulus and I is the second moment of the cross-sectional area, $\frac{1}{12}ab^3$ for a strip a wide and b thick. The equation for the strut is therefore

$$EI\, d^2y/dx^2 = mg(y_0 - y) + F(l - x),$$

or

$$d^2y/dx^2 + \alpha^2 y = \alpha^2 y_0 + (F/EI)l - (F/EI)x,$$

in which

$$\alpha = (mg/EI)^{\frac{1}{2}}. \tag{2.1}$$

For this type of equation, which belongs to the same family as (1.5), the general solution consists of any solution plus the general solution of $d^2y/dx^2 + \alpha^2 y = 0$. Thus a suitable general solution takes the form

$$y = y_0 + (F/mg)(l - x) + A \sin \alpha x + B \cos \alpha x. \tag{2.2}$$

Now at the base, where $x = 0$, $y = y' = 0$. Making these substitutions in (2.2) we find

$$B = -(y_0 + Fl/mg) \quad \text{and} \quad A = F/mg\alpha. \tag{2.3}$$

Fig. 2.5. Notation for the theory of the Euler strut.

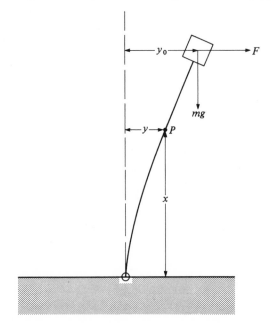

To determine the deflection y_0, we note that there is no bending moment at the top, where $x = l$, so that d^2y/dx^2 must vanish here. From (2.2)

$$0 = \alpha^2 A \sin \alpha l + \alpha^2 B \cos \alpha l,$$

i.e.

$$\tan \alpha l = -B/A = mg\alpha y_0/F + \alpha l. \tag{2.4}$$

Hence the restoring force coefficient μ, for horizontal displacements, which is F/y_0, is found:

$$\mu = mg\alpha/(\tan \alpha l - \alpha l), \tag{2.5}$$

and for the angular frequency of small-amplitude oscillation,

$$\omega = \sqrt{\mu/m} = [g\alpha/(\tan \alpha l - \alpha l)]^{\frac{1}{2}}. \tag{2.6}$$

So long as $\alpha l < \pi/2$, $\tan \alpha l > \alpha l$ and ω is real – the mass oscillates to and fro. But as m is increased so that αl approaches $\pi/2$, $\tan \alpha l$ becomes large and the frequency gets lower. Above the critical mass, $m_c = \pi^2 EI/4gl^2$, at which $\alpha l = \pi/2$, ω^2 is negative and the strut topples over. This analysis does not show whether it finds a new equilibrium (as it does) or whether it collapses completely, either of which is a possibility until a more detailed study resolves the issue.

To see how the frequency approaches zero, write αl as $\pi/2 - z$. Then $\tan \alpha l \sim 1/z$ and $\omega^2 \propto z$. Since for small variations of m around m_c, αl is a linear function of m, we expect a graph of ω^2 against m to fall to zero linearly as $m \to m_c$, as in fig. 2.6. It will be noted that there is no mathematical objection to continuing the line for a while into negative regions of ω^2, to show how the rate of fall from the central position increases with the load. It is well worth while checking this result with the equipment of fig. 2.4.

A quick test of the general correctness of the approach is to fix the critical load m_c with a paper clip and then turn the apparatus upside down, letting the strut hang down over the edge of the table. This is equivalent to reversing the sign of g.

Exercise 2.7. Show that when the strut hangs down

$$\omega = [\alpha g/(\alpha l - \tanh \alpha l)]^{\frac{1}{2}},$$

and that, when the load is critical for the right-way-up position, the upside-down frequency is

$$1.550\sqrt{g/l}, \quad \text{or} \quad T = 4.054\sqrt{l/g}.$$

When I did this experiment, the length of the strut from the base to the centre of the mass was 11.0 cm, and T was 0.435 s. With $g = 9.80 \text{ ms}^{-2}$, a period of 0.430 s would have been expected. As a matter of fact, this agreement to about 1% is somewhat fortuitous, since there are several small corrections to be applied for the finite size of the load. Nevertheless, the general result is satisfactory.

> *Exercise 2.8.* If the apparatus lies on its side, so that the strut is horizontal, what oscillation frequency do you expect when $\alpha l = \pi/2$? You may find it helpful to remember that for small αl, $\tan \alpha l = \alpha l + \frac{1}{3}(\alpha l)^3 + \cdots$ and $\tanh \alpha l = \alpha l - \frac{1}{3}(\alpha l)^3 + \cdots$. When I worked out what I expected and then tested it, I found a significant difference. Do the same for yourself and, if you also find a difference, see if you can explain it. This is a useful object lesson in checking calculations experimentally wherever possible, in case you have made a false assumption.

The phase plane; stability criteria

For the harmonic oscillator, or indeed for any system for which x may be described as a function of x and $\dot{x}$ alone, it is often convenient to represent the behaviour of $x(t)$ as the motion of a point on a two-

Fig. 2.6. Results for the apparatus of fig. 2.4, showing how ω^2 falls linearly to zero with increasing mass, as measured by n, the number of plastic strips added. The periodic time $T = 2\pi/\omega$.

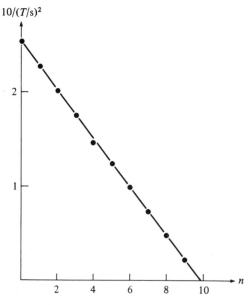

dimensional graph having x and $\dot{x}$ as abscissae and ordinates. To take (1.3) as an example, if we write y for $\dot{x}$,

$$\dot{y} = -(2\lambda y + \omega_0^2 x). \tag{2.7}$$

For every point on the plane (x, y) the y-velocity of the representative point is determined by this equation, while the x-velocity is just y. Hence the direction in which the point moves, and its speed, are precisely specified, and the resulting trajectory on the phase plane shows the behaviour after its start at some chosen (x, y). If we are interested only in the shape of the trajectory we may combine (2.7) with the statement that $y = \dot{x}$ to eliminate t; then

$$dy/dx = -(2\lambda y + \omega_0^2 x)/y, \tag{2.8}$$

thus defining the gradient of the trajectory everywhere. It may be noted that since, for given λ and ω_0^2, dy/dx is unique, the trajectories do not intersect.

It is easy to get a computer to draw the trajectories, but it is more instructive to see how far one can get first without computation. To start with, let us put $\lambda = 0$ (no loss), when the solution of (2.8) is

$$y^2 + \omega_0^2 x^2 = \text{constant},$$

representing a series of similar ellipses (fig. 2.7) with axial ratio ω_0. The representative point obviously (?) goes round clockwise. By writing $y =$

Fig. 2.7. Elliptical trajectories in the phase plane of a harmonic oscillator.

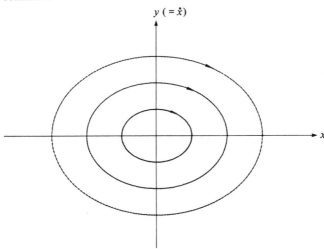

$\dot{x}/\omega_0$ rather than $\dot{x}$, we turn these ellipses into circles. With this new definition of y, (2.8) becomes

$$\mathrm{d}y/\mathrm{d}x = -(x+ky)/y, \quad \text{where } k = 2\lambda/\omega_0. \tag{2.9}$$

Now let k not be zero; then $\mathrm{d}y/\mathrm{d}x = 0$ when $x = -ky$ and $\mathrm{d}y/\mathrm{d}x = \infty$ when $y = 0$. Draw these two lines on the x–y plot and label them 0 and ∞ to indicate that the trajectory passes horizontally through one and vertically through the other. See fig. 2.8(a) where k, representing the lossy

Fig. 2.8. (a) An exercise in sketching (broken curve) the solution of (2.8); (b) a computed trajectory.

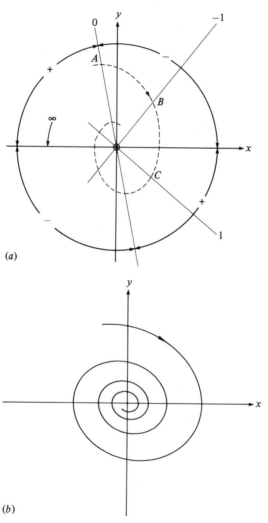

term, is relatively small (0.2). Further, when x and y are both positive, dy/dx is negative; by such considerations we may divide the diagram into four regions, labelled by the sign of the gradient. Given this meagre information, we may make a shot at drawing the curve, starting with a horizontal tangent at A (and noting, if we feel like it, that it crosses the y-axis at a gradient of $-k$ – sloping down by as much as the 0-line slopes back). A critical examination of the result might involve noting that the slope is -1 when $y = x/(1-k)$, and $+1$ when $y = -x(1+k)$. I drew the curve shown without benefit of this last remark. When the lines for ± 1 are added it becomes clear that the slope is too great both at B and C; thus the true curve is probably somewhat fatter, and this may imply that the decrement per cycle is less than shown. The computed curve (fig. 2.8(b)) confirms this.

Much of this, however, is unnecessarily pernickety – the important point, to be realized confidently with only the 0 and ∞ lines, is that we have a spiral trajectory, traversed in such a sense that the oscillation is damped. If that is enough for our needs – and it very often is – well and good; but if you want much more, you will either have to draw very carefully or use a computer.

In this example the focus, or stable equilibrium position, lies at the origin, which in general is not necessary. The characteristic of a point of stable or unstable equilibrium is that there $\dot{x}$ and $\dot{y}$ both vanish. Hence if dy/dx is expressed as $f_1(x, y)/f_2(x, y)$, it is a point at which both f_1 and f_2 vanish, i.e. an intersection of 0 and ∞ lines. In the following exercises, find all the points of equilibrium, and pay particular attention to what happens in their vicinity before attempting to draw complete trajectories.

Exercise 2.9. Sketch typical trajectories for the following differential equations:

(a) $dy/dx = -(3x + 5y - 15)/(7x - 2y - 14)$. You may have difficulty with the distant behaviour. Don't be ashamed to resort to simple analysis to find what happens. There is one saddle-point.

(b) $dy/dx = (12x + y)/(x - \sin y)$. There are five foci and four saddle-points.

In the neighbourhood of any equilibrium point, at which both $\dot{x}$ and $\dot{y}$ vanish, $\dot{x}$ and $\dot{y}$ are to first-order linear functions of the distance from the equilibrium point. Thus in Ex. 2.9(a), the equilibrium point, where $3x + 5y - 15 = 0$ and $7x - 2y - 14 = 0$, has co-ordinates $x = 100/41$ and

$y = 63/41$; changing the origin to this point by putting $X = x - 100/41$, $Y = y - 63/41$, we find

$$\mathrm{d}Y/\mathrm{d}X = -(3X + 5Y)/(7X - 2Y),$$

with the constants eliminated but the coefficients unchanged, as expected for linear numerator and denominator. In general, Taylor expansion of $f_1(x, y)$ and $f_2(x, y)$ close to the equilibrium point, where the linear terms give a good enough approximation, allows one to write

$$\mathrm{d}Y/\mathrm{d}X = (X\,\partial f_1/\partial x + Y\,\partial f_1/\partial y)/(X\,\partial f_2/\partial x + Y\,\partial f_2/\partial y),$$

the differential coefficients being taken at the equilibrium point. If you do not mind simple numerical solution of trigonometrical equations, find the nine equilibrium points in Ex. 2.9(b), at which $x + \sin 12x = 0$, and show that the nine linearized forms of $\mathrm{d}Y/\mathrm{d}X$, valid only near the equilibrium points, all have $12X + Y$ as numerator, and denominators $X - Y$ (once), $X + 0.9583Y$ (twice), $X - 0.8763Y$ (twice), $X + 0.4859Y$ (twice) and $X - 0.3292Y$ (twice).

Reverting to the time-varying representation, we can say that in this linearized form, in general

$$\dot{X} = \alpha X + \beta Y,$$

and   (2.10)

$$\dot{Y} = \gamma X + \delta Y$$

α, β, γ and δ being constant coefficients. To see the range of possible behaviours generated by these equations, note that Y may be eliminated to give

$$\ddot{X} - T\dot{X} + \Delta X = 0,$$ (2.11)

where T is the trace, $\alpha + \delta$, and Δ the determinant, $\alpha\delta - \beta\gamma$, of the matrix

$$\begin{bmatrix} \alpha & \beta \\ \gamma & \delta \end{bmatrix}.$$

Clearly, then, to find the behaviour of the trajectory near the equilibrium points, we need only calculate T and Δ, and compare (2.11) with the damped harmonic oscillator equation (1.3), setting $-T/2 = \lambda$ and $\Delta = \omega_0^2$. Then it follows immediately that

(1) if $\Delta > \frac{1}{4}T^2$, the solution is oscillatory, decaying when T is negative (*focal stability*) and growing when T is positive (*focal instability*).

(2) if $0 < \Delta < \frac{1}{4}T^2$, the solution is non-oscillatory, with *nodal stability* for T negative and *nodal instability* for T positive.

(3) if $\Delta < 0$, the system always shows *saddle-point instability*.

The ranges are shown in fig. 2.9, together with characteristic trajectories.

Exercise 2.10. Reverting to Ex. 2.9(a), take $\dot{x} = 7x - 2y - 14$, $\dot{y} = 15 - 5y - 3x$, and determine the character of the equilibrium point. Do the same for the nine equilibrium points in Ex. 2.9(b), for which relevant data are given below the exercise. Take $\dot{Y} = -(12X + Y)$ for each. Compare with your sketched solutions, not neglecting the directions of the arrows, which are determined once dY/dX has been dissected into $\dot{X}$ and $\dot{Y}$.

The above analysis does not apply only to cases where $y = \dot{x}$, but to any system fully determined by two variables, and having an equilibrium configuration, whether stable or unstable.

Exercise 2.11. The two LR circuits of fig. 2.10 are inductively coupled as shown. Write expressions for di_1/dt and di_2/dt in terms of i_1 and i_2. Show that

$$\Delta = R_1 R_2/(L_1 L_2 - M^2) \quad \text{and}$$

$$T = -(L_2 R_1 + L_1 R_2)/(L_1 L_2 - M^2).$$

Now observe that for any coupled pair of inductors, $L_1 L_2 \geqslant M^2$. Hence show that the system shows nodal stability when both R_1 and R_2 are positive, nodal instability when both are made negative, e.g. by use of the circuit in fig. 2.2, and saddle-point instability if R_1 and R_2 are opposite in sign.

Fig. 2.9. Domains of the various types of stability and instability for the general linear equations in two variables, (2.10); $\Delta = \alpha\delta - \beta\gamma$, $T = \alpha + \delta$. Typical trajectories near the stationary points are shown.

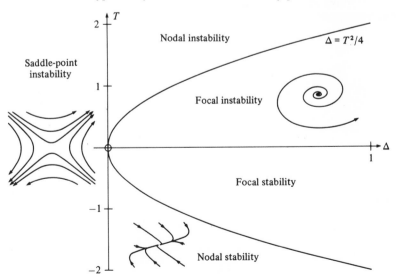

A second example is very important in dynamics, and is concerned with the stability of a rigid body spinning freely in space.[2] A rigid body possesses an inertia tensor I_{ij} which is symmetrical, and all we need to know is that there can always be found three orthogonal axes in the body with respect to which the tensor is diagonal; there are thus three principal moments of inertia, I_1, I_2 and I_3. When the body is spinning about any one of these axes, its angular momentum vector L also points along this axis, and if the body continues spinning freely like this both ω, the angular velocity, and L remain fixed in space. Since L must be conserved in free motion, any one of those three motions, rotation about a principal axis, can remain unchanged. The question is, what happens if the body is not spinning about a principal axis, and in particular if ω is displaced slightly from a principal axis is the motion stable? It can be seen that if the three principal values of I are not the same, L will not be parallel to ω, for $L_i = I_{ij}\omega_j$. If then the body simply continued to spin about ω, fixed in space, L would spin with it and contradict the conservation law. In fact ω must continually migrate relative to the principal axes of the body, and must point appropriately in space so that L is invariant. The rules governing the migration of ω relative to the body are Euler's equations:

$$I_1\dot{\omega}_1 = (I_2 - I_3)\omega_2\omega_3$$
$$I_2\dot{\omega}_2 = (I_3 - I_1)\omega_3\omega_1 \qquad (2.12)$$
$$I_3\dot{\omega}_3 = (I_1 - I_2)\omega_1\omega_2$$

Exercise 2.12. If ω lies close to a principal axis, e.g. if ω_2 and ω_3 are much less than ω_1, $\dot{\omega}_1$ as given by the first equation is a second-order small quantity. To study the motion in these circumstances, take ω_1 as constant and write down the linear equations relating $\dot{\omega}_2$ and $\dot{\omega}_3$ to ω_2 and ω_3. Hence show that $T=0$ and $\Delta = -\omega_1^2(I_3 - I_1)(I_1 - I_2)/I_2 I_3$. The significance of $T=0$ is that there are no frictional forces, so that if Δ is positive ω_2

Fig. 2.10. Coupled *LR* circuits for Ex. 2.11.

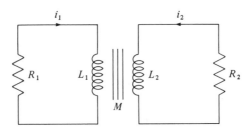

and ω_3 will undergo small oscillations about zero without damping. Show that ω_2 and ω_3 oscillate in phase quadrature to cause ω to execute an elliptical path around the nearest principal axis, with axial ratio $[I_2(I_1 - I_2)/I_3(I_1 - I_3)]^{\frac{1}{2}}$ and period

$$\frac{2\pi}{\omega_1}[I_2I_3/(I_1 - I_2)(I_1 - I_3)]^{\frac{1}{2}}.$$

Under what conditions is Δ positive? Show that if I_1 is either the largest or the smallest of the three Is, the motion is precessional, while if I_1 has the intermediate value the system shows saddle-point instability.

The behaviour can be demonstrated with the aid of the sphere shown in fig. 2.11 which spins freely on an air cushion. This demonstration requires skilled craftsmanship, for the sphere, 8″ in diameter, has a core in the form of a steel block 6″ × 3″ × 2″. This is embedded in a stack of 1″-thick blocks of bakelized paper and the whole is accurately turned spherical before being painted matt black, to take chalk. Compressed air is supplied to the cup and the sphere is then spun slowly by hand. A piece of chalk lightly applied shows how ω migrates. At any given instant ω lies at the centre of curvature of the arc traced out by the chalk. The inferred trajectory of ω is shown as a broken line drawn on the photograph. For comparison, fig. 2.12 shows solutions of Euler's equations plotted on a sphere to show how X, the largest I, and Z, the smallest I, are foci of stable motion, while Y, the intermediate axis of inertia, cannot maintain steady motion. The parameters used in this calculation (Is in the ratio 1.248:1.211:1) are those measured for the sphere by means of a torsional pendulum.

Linear systems of higher order; Routh–Hurwitz and Nyquist

When the state of a system is defined by more than two variables the phase plane must give way to a phase space of as many dimensions as there are independent state variables. The behaviour is still described by a line trajectory, but obviously a much more elaborate menu of possibilities must now be expected. Often enough, however, one is not greatly concerned how a system approaches, or diverges from, any state of equilibrium; all that matters is whether it is stable or unstable. For the simple systems we have discussed, whose second-order linearized equations admit of two exponential solutions, $e^{p_1 t}$ and $e^{p_2 t}$, it is necessary and sufficient for stability that the real parts, Re $[p_1]$ and Re $[p_2]$, shall both be negative. Otherwise a solution with positive real part will grow and destroy equilibrium.

The same criterion must obviously hold when the linearized equations of motion lead to a differential equation with constant coefficients, but of order higher than two. The solutions will still be of the form e^{pt}, but there will be more, as many as the order of the equation, and the real part of every p must be negative. Thus the differential equation for one of the variables may take the form

$$a_n \, d^n x/dt^n + \ldots + a_1 \, dx/dt + a_0 x = 0, \tag{2.13}$$

Fig. 2.11. The 8″-diameter sphere rotates freely on its air cushion, supplied from a compressed air cylinder. Embedded at its centre is a steel block, $6'' \times 3'' \times 2''$. The serpentine line is made by holding a piece of soft chalk lightly against it as it spins; the broken line, added afterwards, indicates the trajectory along which the axis of rotation has migrated. It should be compared with one of the lines in fig. 2.12. The square indicates the axis of I_{max}, the circle I_{min} and the triangle the intermediate axis.

in which all a_n are real. On substituting $x = e^{pt}$, we find p must satisfy

$$a_n p^n + \ldots + a_1 p + a_0 = 0, \qquad (2.14)$$

from which the n independent values of p emerge (if there are degenerate solutions having the same p, they must be treated separately in writing down the general solution, but they do not affect the question of stability). It is a matter for advanced algebraic analysis to determine the conditions under which all p have negative real parts, but the result can be written down as a set of rules, the Routh–Hurwitz criteria,[3] which involve subjecting (2.14) to a systematic procedure, as follows.

First verify that all the a_n are positive; if they are not, proceed no further – the system is unstable. If they are all positive, stability is not guaranteed, and the next stage may prove laborious.

Calculate a set of numbers (the *test determinants*) defined thus:

$$T_1 = a_{n-1}$$

$$T_2 = \begin{vmatrix} a_{n-1} & a_n \\ a_{n-3} & a_{n-2} \end{vmatrix}$$

$$T_3 = \begin{vmatrix} a_{n-1} & a_n & 0 \\ a_{n-3} & a_{n-2} & a_{n-1} \\ a_{n-5} & a_{n-4} & a_{n-3} \end{vmatrix}$$

Fig. 2.12. Calculated trajectories for ω in the sphere of fig. 2.11. The labels are the same, the axis (x) of maximum moment of inertia being marked ■, while that of minimum moment (z) is ●. Both are stable (or, in the absence of friction, stationary in small encircling orbits, analogous to states of neutral equilibrium). There is no encircling orbit for the intermediate axis (y), shown as ▲, which is a saddle-point, with and without friction.

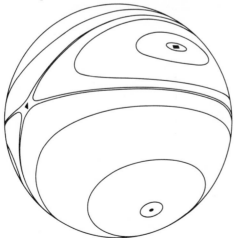

and so on. The entry at the top left is always a_{n-1}, and the subscripts increase by 1 in going a step to the right, and decrease by 2 for a downwards step. Any term with a subscript greater than n or negative is replaced by zero, and T_r is an $r \times r$ determinant. There are as many test determinants as the order of the equation, and *if all T_r are positive, the system is stable.*

> *Exercise 2.13.* The feedback-controlled amplifier indicated inside the broken lines in fig. 2.13 may be considered as a black box having the properties that $V_Q = AV_P$, the input impedance at P is infinite, and the output impedance at Q zero. Taking q as the charge on one capacitor plate, solve the circuit to show that
>
> $$(LL_0 - M^2)\dddot{q} + (R_0L + RL_0)\ddot{q}$$
> $$+ (MA/C + RR_0 + L_0/C)\dot{q} + (R_0/C)q = 0.$$
> $$(2.15)$$
>
> Hence show that if M is positive the circuit is always stable, i.e. does not oscillate, while if M is negative it becomes unstable when A is increased above the value A_c such that
>
> $$|MA_c| = CRR_0 + L_0 - R_0(LL_0 - M^2)/(R_0L + RL_0).$$

So far, the Routh–Hurwitz test only determines that the circuit will become unstable when $A > A_c$, but it does not give the form of the instability. If, however, A is put equal to A_c in (2.15) the equation takes the form

$$a\dddot{q} + b\ddot{q} + (ca/b)\dot{q} + c = 0,$$

Fig. 2.13. A simple maintained oscillator circuit, for analysis in Ex. 2.13.

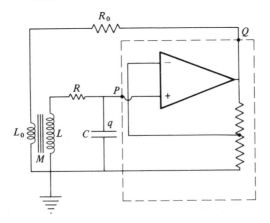

where $a = (LL_0 - M^2)$, $b = (R_0L + RL_0)$ and $c = R_0/C$. This equation divides into two equivalent parts, $a\ddot{q} + (ca/b)\dot{q} = 0$ and $b\ddot{q} + c = 0$, indicating a solution which describes steady oscillation at $\omega = (c/b)^{\frac{1}{2}} = [R_0/C(R_0L + RL_0)]^{\frac{1}{2}}$. If L_0 is small and A correspondingly large, $\omega \sim (LC)^{-\frac{1}{2}}$, the natural frequency of the resonant circuit. It is intuitively clear, if not mathematically proved, that when A is slightly larger than A_c the state $q = 0$ has focal instability, oscillating with growing amplitude.

An alternative approach to the stability of a complex system is by way of *Nyquist's theorem*,[4] which is best appreciated by describing a specific application. The circuit of fig. 2.13 is essentially a feedback loop running from Q through M and the resonant circuit to P and so, via the amplifier, back to Q. Let us now break the loop at some convenient point, P say, as in fig. 2.14, and ask what oscillatory e.m.f. $v'e^{i\omega t}$ appears at P' when $ve^{i\omega t}$ is applied at P. The ratio v'/v is the loop gain $G(\omega)$, a function of ω and in general complex.

Exercise 2.14. Solve the circuit to show that
$$G = (AM/C)/[(R_0 + i\omega L_0)(R + i\omega L - i/\omega C) + \omega^2 M^2].$$
(2.16)

When $G(\omega)$, for real ω, is plotted on the complex plane, each value of ω contributes a point to a continuous trajectory. It takes some little practice to sketch $G(\omega)$, and we shall not go into this matter here. Much can be learnt by general argument. For instance, $G(0)$ and $G(\infty)$ in (2.16) are both zero – the trajectory starts and stops at the origin. In general, $G(\infty) = 0$, but it is not necessary for $G(0)$ to be zero; it is the presence of M in the loop that leads to this here. If negative ω are also included, $G(-\omega) = G^*(\omega)$, and the trajectories for positive and negative ω are mirror images in the real axis. The complete trajectory, for $-\infty < \omega < \infty$, then starts and finishes at 0 and forms a symmetrical closed curve.

When A is small the excursion from zero is correspondingly small, and the whole trajectory scales as A. For some value or values of A, A_c say, the trajectory will pass through the point $(1, 0)$ at which G is real and equal to unity. At the frequency ω_c where this happens, P and P' vibrate exactly together, and may be joined without altering the behaviour, which is steady vibration at frequency ω_c. If A is now increased so that the trajectory encircles $(1, 0)$, the vibration at this frequency begins to grow, and the circuit has become unstable. This exhibits (but, of course, does not prove) a very general property of the *Nyquist diagram*, as the trajectory of $G(\omega)$ is called – if it encircles the point $G = 1$, the system will be unstable when the feedback loop is connected.

Exercise 2.15. Find A_c for the circuit. Note that if $G(\omega)=1$ the denominator of (2.16) must be real. Using this fact you should verify without difficulty both the critical value of A and the frequency at which the oscillation begins to grow as A is raised above A_c, as found by the Routh–Hurwitz procedure.

With the circuit of fig. 2.14, the Nyquist test of stability is probably easier than the Routh–Hurwitz; if, however, the differential equation is of higher order, say 6, Routh–Hurwitz, though tedious, involves no further thought, while plotting the Nyquist trajectory may prove tricky. This apart, the Nyquist test has two major advantages:

(a)   it requires no analytical formula and can be used with experimental measurements of the open loop gain $G(\omega)$, and

(b)   it applies even when the system is not described by a differential equation of finite order.

Such a circumstance occurs whenever there is a time delay in the loop, so that an input $f(t)$ emerges at $f(t-t_0)$.

A model of delayed feedback

To illustrate this point, consider the water mobile shown in fig. 2.15. A fan of water, W, falls on a vane V, hinged at H so that the amount of water falling into the channel C can be varied by pulling on the string S. The water takes time t_0 to run down the channel and into the bucket B; B has a hole in the bottom and is pivoted and counterpoised on a horizontal arm A, to which the string S is attached. For convenience let us adjust the

Fig. 2.14. The same as fig. 2.13, but with the feedback loop opened at PP'.

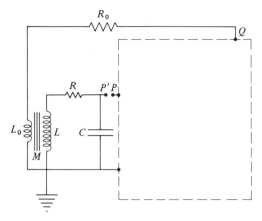

mobile so that when V is vertical the steady stream of water keeps A horizontal with a height h_0 of water in the bucket. We have now to enquire whether this state of affairs represents stable equilibrium. At first sight it does, since if too small a flow makes B rise, V moves to the right, the water flow increases and B falls again with the increased head of water in the bucket. This is a typical case of negative feedback, designed to stabilize the system, but we have reckoned without the time delay t_0.

Now let us write down some equations, first cutting the string at Z to open the feedback loop. Let the displacement of the right-hand part of the string be z, and of the left-hand part z', so that the open loop gain $G(\omega)=z'/z$. If now $z=a\mathrm{e}^{i\omega t}$, where a is small, the water will flow into the top of the channel at a rate $V_0+a\phi\mathrm{e}^{i\omega t}$, where V_0 is the equilibrium flow rate and ϕ the flux per unit length of the fan. What comes out at the bottom of the channel, delayed by t_0, is $V_0+a\phi\mathrm{e}^{i\omega(t-t_0)}$.

Next we turn to the bucket on its pivoted arm, and for simplicity we shall suppose that the arm is damped and takes up its equilibrium position instantaneously, thus following exactly the weight of water in the bucket. This we can formally express by saying that the bucket falls ph when the head of water increases from h_0 to h_0+h. And if the rate at which water

Fig. 2.15. The water mobile, illustrating delayed negative feedback, as described in the text.

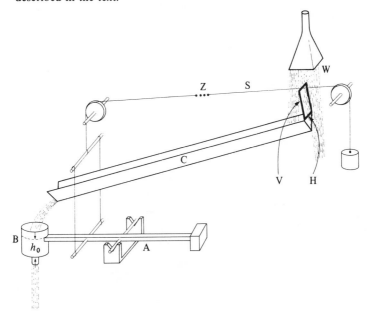

flows out of the bucket is proportional to the head, we can define a time constant τ for emptying the bucket by the equation

$$\mathrm{d}h/\mathrm{d}t = -(h_0 + h)/\tau.$$

But this is to omit the flow of water in, and the complete equation of motion is

$$\mathrm{d}h/\mathrm{d}t + (h_0 + h)/\tau = [V_0 + a\phi e^{i\omega(t - t_0)}]/A$$

where A is the cross-sectional area of the bucket. Since V_0 alone keeps the bucket with a head h_0, the steady-state terms cancel, leaving an equation whose solution is

$$h = [(a\phi\tau/A)e^{-i\omega t_0}/(1 + i\omega\tau)]e^{i\omega t}.$$

Finally we note that the bucket falls ph, and if S is attached a fraction α of the distance from pivot to bucket, $z' = -\alpha ph$. Hence

$$G(\omega) = z'/z = \beta[-e^{-i\omega t_0}/(1 + i\omega\tau)], \tag{2.17}$$

where $\beta = \alpha p\phi\tau/A$. The overall amplification is controlled by β, which can be varied by changing α, i.e. the point of attachment of S. It is the quantity in square brackets that we plot to obtain the Nyquist diagram.

Now on the complex plane $1 + i\omega\tau$ is a straight line passing through 1 and parallel to the imaginary axis ($\mathscr{L}_1$ in fig. 2.16). Its reciprocal is the circle $\mathscr{L}_2$. The effect of $e^{-i\omega t_0}$ is to rotate each point appropriately round the origin, the point shown as $\omega = 0$ remaining fixed, while towards $+\infty$ the point suffers many clockwise turns, and towards $-\infty$ many anticlockwise turns, to give a spiral trajectory as in fig. 2.17, in which the positive range of ω is shown, for $t_0 = 2\tau$. From the crossing point for the largest turn it may be seen that the system is stable so long as $\beta < 1.52$. This is confirmed

Fig. 2.16. The trajectory $r = 1 + i\omega\tau$ $(-\infty < \omega < \infty)$ denoted by $\mathscr{L}_1$, and its reciprocal, $\mathscr{L}_2$, a circle centred at $-\frac{1}{2}$.

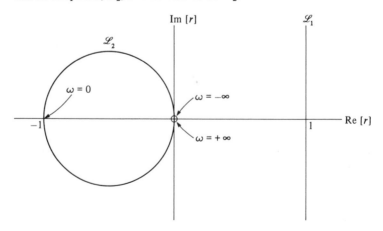

by computation of the solution of the equations of motion when the loop is closed so that $z' = z$. The very nearly steady oscillation of fig. 2.18(a) was obtained for $\beta = 1.50$. When β is somewhat less the oscillations die away gradually (focal stability), when it is larger they grow (focal instability). At the critical value of β, the oscillation period is $2.75t_0$ (this figure depends, of course, on the chosen ratio, t_0/τ).

The remaining curves were obtained on the assumption that the water fan had a sharp edge, and that stops prevented the vane from moving more than a short distance beyond the edges. As the oscillation grows,

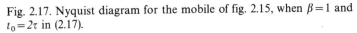

Fig. 2.17. Nyquist diagram for the mobile of fig. 2.15, when $\beta = 1$ and $t_0 = 2\tau$ in (2.17).

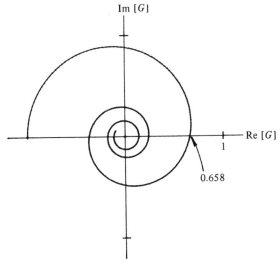

Fig. 2.18. Oscillatory response of the mobile for various values of β near and above the critical value 1.52: (a) $\beta = 1.5$, (b) 1.7, (c) 3, (d) 10.

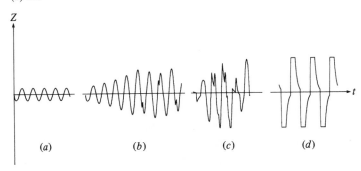

eventually the vane travels beyond the fan edges, and there are gaps in the flow of water down the channel. When these gaps reach the bucket it has no option but to relax exponentially towards its equilibrium position, and this accounts for the jags in the motion. When $\beta = 3$ (curve (c)) the behaviour is chaotic, but larger values (e.g. 10, as in (d)) restore regular motion. Now, however, the fan spends much of the time against the stops. Suppose it rests against the outside stop, so that a full surge of water starts down the channel; after t_0 it arrives at the bucket which falls precipitately, cutting off the flow. Now there is a lull while the bucket empties itself (the flat and exponential parts of the cycle) but because no water arrives it rises so that the fan moves to restore the full flow, and the cycle repeats itself.

This model has illustrated, among other things, how ideas of stability and tests such as Nyquist's can be applied to systems quite other than electrical circuits. Indeed, the general theory of control,[5] of which Nyquist's theorem provides an example, is applicable to a very wide range of problems in physics, engineering, physiology and economics – anywhere, in fact, where the running of the system is controlled by some form of feedback. This field is so large, so important and so technically involved that I shall not pursue it further, but refer you to standard texts.

Further comments on delayed feedback

The observation that at critical feedback the period of oscillation is proportional to t_0 indicates that delay is necessary for oscillation to occur. This is indeed obvious from the equation for the closed loop system, which takes the form

$$\tau \dot{z} + z + \beta[z] = 0 \tag{2.18}$$

in which $[z]$ is the delayed value, $[z(t)] = z(t - t_0)$. If $t_0 = 0$, $[z] = z$ and $\tau z + (1 + \beta)z = 0$, having a simple exponential decrement with time constant $\tau/(1 + \beta)$. The more the feedback, the faster the response (note that positive $\beta \equiv$ negative feedback).

Let us now proceed to find the general solution of (2.18) which, since the equation is linear, can be written as a sum of exponentials of the form e^{pt}. Substituting we find

$$1 + P + \beta e^{-rP} = 0, \tag{2.19}$$

where $P = \tau p$ and $r = t_0/\tau$.

> *Exercise 2.16.* (a) Show that when $r = 2$ the largest value of β for which (2.19) has real solutions for P is $\frac{1}{2}e^{-3}$, i.e. 0.025; and that for smaller positive values of β there are two negative values of P. In this narrow range of negative feedback the system is in nodal stability, like an overdamped harmonic oscillator; unlike the

latter, however, the product of the two values of P is not constant. The behaviour for small β is shown in the inset to fig. 2.19.

(b) When $\beta > \frac{1}{2}e^{-3}$, the solutions are oscillatory, and P is complex. Write $P = P' + iP''$, and show that

$$-[(1+P')+iP'']e^{irP''} = \beta e^{-rP'}.$$

The right-hand side is real and positive, and therefore the phase angles of $-e^{-irP''}$ and $(1+P')+iP''$ are equal. Hence, show that

$$P' = -(1+P'' \cot rP'') \tag{2.20}$$

and that

$$\beta = P''e^{rP'} \operatorname{cosec} rP''. \tag{2.21}$$

It is now easy to compute how P', representing the decrement $(P'<0)$ or amplification $(P'>0)$ of the oscillations, depends on β, as is shown in fig. 2.19. Only the first five independent modes, out of an infinite number, are shown. Each decays rapidly when β is small, and each crosses the axis, to turn into a growing mode, at a critical value of β, as shown, which causes one of the spiral turns in fig. 2.17 to pass through the point 1. The frequencies P'' are not shown, but are readily found from (2.20) by putting $P'=0$: if $r=2$, $\omega = 1.1244$, 2.543, 4.408, etc.

Fig. 2.19. To show successive normal modes of the mobile passing from negative (decaying) values of P' to positive (increasing) as β is raised. The inset shows the region in the broken rectangle enlarged horizontally, to exhibit the bifurcation of the lowest mode from focal to nodal stability as $\beta \to 0$.

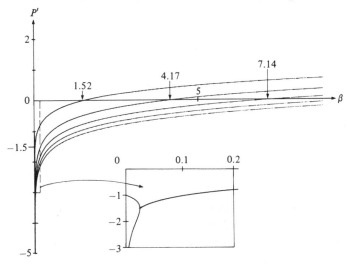

Only the lowest frequency shows critical damping as $\beta \to 0$; the other curves in fig. 2.19 run down to $-\infty$. At all values of β there is an infinite number of independent exponential solutions, though most are very highly damped, and the general solution of (2.18) is an arbitrary sum of these. The strong contrast with a harmonic oscillator is readily understood when one realizes that the latter's initial state is completely specified by the displacement and velocity, $x(0)$ and $\dot{x}(0)$, so that only two independent solutions suffice. With (2.18), however, the whole initial pattern of z already in the channel must be specified, since it is this that controls the behaviour in the first t_0 of the operation; an infinite number of functions is needed for this purpose.

The advantage of strong negative feedback in hastening the approach to equilibrium is soon lost once delay is introduced. Long before β has reached its critical value the decay of oscillation has become slow enough to offset the initial quick response, and the longer the delay, the less the tolerable feedback.

Political implications

[When one leaves the realm of the exact sciences and technologies to apply mathematical arguments to biological, sociological or political problems, it becomes more essential at each step to replace the largely unmeasured complexities of the real world by models simple enough to allow analysis and computation. Often enough, observation will show the strengths and weaknesses of the model, so that it can either be refined or used with caution, bearing in mind that its long-term (or even short-term) predictions are likely to be astray. Provided, however, some degree of correlation is observed between model and reality, the model may prove very valuable. Rarely will it serve to make firm predictions such as physicists (alone among natural scientists) make with confidence. On the other hand, the model may open our eyes to possibilities of behaviour which otherwise would have been quite overlooked. The disastrous effects of delay on negative feedback is a case in point, and I propose to examine briefly the possible application of our calculations in government.

One of the functions of benign government (that is, one that considers the interests of the ruled above those of the rulers) is to ensure, as far as possible, that the life of the country remains reasonably well balanced. To take a particular example, an industrial country depends on its educational system to produce not only educated, responsible citizens, but as well such a supply of trained specialists as are needed to run, improve and make competitive the wealth-producing industries. If there are too few, as has happened in Britain at times, government may choose to offer incentives to cause more talented youth to embark on a career in technology. This is an application of negative feedback, to observe an imbalance and act so as to eliminate it. In this case, however, there is an inevitable delay of many years between the act, which initiates novices into the training process, and the useful outcome in the form of trained engineers. Our model leads us to believe that too decisive government action might be counter-productive – too many students starting would have no effect for some years, and then would saturate the market. Not only that, but for years to come the students already in the pipeline would pour out into a jobless arena. At which point, a government that had not yet learned its lesson would discourage any more students from entering on engineering courses, and in due time a famine of trained young people would once again paralyse industry and lead to demands for corrective action. Now in fact this has not happened, not (I think) because governments were alert to, and swayed by, this view of the problem, but because the issue has never been a vote-catcher such as would lead to unthinking and damaging over-reaction.

There are, however, public issues of far greater sensitivity where politicians are under the strongest pressure to provide instant cures for perceived ills, and of these the state of a country's economy is the most serious, especially as measures to adjust the economy cannot take effect immediately. We have here an ideal situation for delayed negative feedback, and it is not unreasonable to ascribe some of the fluctuations in the economic state of Britain to a too-vigorous application of feedback. It is a curious, and rather disheartening, paradox that matters of limited interest are likely to run on an even keel while the most important matters are caused to pursue a bucketing course through excess of zeal.

It has not always been so, because before the advent of television and the myriad technical advances that have sprung from the transistor, the public were largely ignorant of the information that now pours over it from every side. Moreover, communication was slow, and the nation at large lacked the means to respond to every new revelation with concerted demands for corrective action. The delays of implementation were present then as now, but the public demand, meaning the strength of negative feedback demanded, was less insistent. It is unfortunately necessary to recognize that well-meaning belief in public participation in the political process is likely to prove a dangerous delusion, springing from a failure to appreciate the potential for instability in a modern democracy. This is not to say that we should welcome dictatorship – that is to jump from the frying-pan into the fire. But there is a very real need to try to restore the public's faith in their representatives, and to leave them alone much more to do what they can to solve problems at the appropriate speed. How this is to be done is beyond my imagination, but it is certainly possible if the generality of people wish it – how can we be encouraged to realize that the kitchen can tolerate only so many cooks, and that we should do better to seek out good cooks and look forward, at the very least, to sound, plain food?]

3

Response of linear systems

In the last chapter we saw how to verify that the equilibrium state of a system is stable. Let us now take for granted that we are dealing with a stable system, and discuss how it will respond when gently disturbed, so as not to arouse the non-linearities that every system shows when attacked sufficiently strongly. These are the subject of the next chapter.

There is no need in the first instance to specify the system in detail. All that is needed is a specification of the input and output functions, f_{in} and f_{out}. The input $f_{in}(t)$ is the applied disturbance and the output $f_{out}(t)$ the resulting response. With an engineering structure f_{in} might be an applied force, e.g. a load at some point on a bridge, while f_{out} might be any resulting displacement, of the roadway or of a pier. The form of $f_{out}(t)$ may depend strongly on the form of $f_{in}(t)$ – in the probably mythical story of the regiment who broke a bridge by marching in step across it, a steady force f_{in} would have produced only a small steady deflection f_{out}, while an oscillatory force, presumably at the resonant frequency, supposedly caused a large oscillatory deflection, so large that the linear range of behaviour was exceeded to the point of collapse. Alternatively, the input may be an electric or magnetic field applied to a solid body, and the output the resulting electric or magnetic moment. As nuclear resonance[1] shows, an appropriate input, at the resonant frequency, may lead to a strong response, while at almost any other frequency hardly anything happens. On the other hand, in many dielectrics the response is only a slowly varying function of frequency. For our analysis it does not matter in the least why the relation between f_{in} and f_{out} takes the form that is observed, or whether its theoretical calculation involves only the certainties of classical mechanics (as for the bridge) or the probability distributions of quantum mechanics (as for magnetic materials and nuclear resonance). In so far as a measured output is uniquely related to a measured input, we are dealing with macroscopic phenomena. There are many reasons why repetitions of the same trial should yield different results, but we cannot talk of a unique relation between input and output until enough measurements have been made to average out fluctuations.

A linear system has the property that the output scales with the input – if a time-dependent input $f_{in}(t)$ produces an output $f_{out}(t)$, then $\alpha f_{in}(t)$ produces $\alpha f_{out}(t)$. Let us write this in shorthand notation,

$$\text{if } f_{in} \rightarrow f_{out}, \quad \text{then } \alpha f_{in} \rightarrow \alpha f_{out}.$$

Moreover, the principle of superposition applies:

$$\text{if } f_{in} \rightarrow f_{out} \quad \text{and} \quad f'_{in} \rightarrow f'_{out},$$
$$\text{then } f_{in} + f'_{in} \rightarrow f_{out} + f'_{out}.$$

For example, an oscillatory force $A \cos \omega t$ might cause a mechanical system to vibrate as $B \cos(\omega t + \phi)$, while a steady force F might cause a steady displacement D. Then doubling either force will double the resulting vibration or steady displacement, and if they are both applied together the system will vibrate as before, but about a new mean position,

$$A \cos \omega t + F \rightarrow B \cos(\omega t + \phi) + D.$$

In this example there is already an implied specialization, that the system possesses the property of *time-invariance*. The assumption that a steady force produces a steady displacement implies that the system being acted on by the force is not changing with time. A ball of radius a sinking slowly through an oil of high viscosity has a velocity $v \ (=f_{out})$ proportional to its (weight – upthrust), $W \ (=f_{in})$, in accordance with Stokes's law, $W = 6\pi\eta av$; so long as the oil remains at constant temperature $v \propto W$, with a time-independent constant of proportionality. This is typical of a linear system, and linearity is not destroyed if the temperature is steadily rising and η correspondingly falling; at any instant $v \propto W$, but now the constant of proportionality is time-dependent. Again, a resonant system acted on by an oscillatory force at constant frequency may vibrate in synchronism with only a small amplitude, and in antiphase, if its natural frequency is lower than that of the applied force. But if the spring which supplies the restoring force is slowly made stronger, to raise the natural frequency, the vibration will increase, pass through a large amplitude around the moment of resonance and then die away again; meanwhile the phase of vibration will have shifted by nearly π so as to be nearly in phase with the driving force. There is nothing in this quite complicated pattern in time to violate linearity – an applied force having two components at different frequencies would elicit a response that went through two resonant peaks, as if each component acted alone and their combined effect was simply the sum of the separate effects.

A simple resonant system is described by the differential equation (cf. (1.3))

$$a_0 x + a_1 \dot{x} + a_2 \ddot{x} = F(t),$$

where the coefficients do not contain x or any of its derivatives. In this form, F plays the part of f_{in} and x the part of f_{out}. You will easily satisfy yourself that superposition holds – if $x_1(t)$ is a solution when $F = F_1(t)$, and $x_2(t)$ when $F = F_2(t)$, then $x_1 + x_2$ is a solution when $F = F_1 + F_2$. It is not necessary to assume that a_0, a_1, etc., are time-independent. If they are, the solution is the standard solution for a resonant system with fixed parameters, but in the example in the last paragraph we allowed a_0 to be $a_0(t)$, without spoiling the linearity.

Impulse response function

In what follows I shall have nothing to say about time-dependent systems, which generally need special treatment for each separate case. But there is no need to restrict the form of the time-independent system except to ensure linearity of response. Thus we may consider systems described by differential equations of any order,

$$a_0 x + a_1 \dot{x} + a_2 \ddot{x} + a_3 \dddot{x} + \ldots = F(t), \tag{3.1}$$

only demanding that the coefficients be strictly constant. This is, however, not the most general form describing linear, time-independent response. The water mobile analysed in the previous chapter is a system requiring a   31 differential equation of infinite order; nothing less can describe a time-delay $f_{out}(t) \propto f_{in}(t - t_0)$, but luckily there are better ways. The most general form can be derived with the help of the superposition principle. Any input $f_{in}(t)$ may be regarded as a continuous sequence of short impulses, e.g. in the interval t to $t + \delta t$ the input is an impulse of strength $\bar{f}_{in}(t)\,\delta t$, $\bar{f}_{in}$ being the average over the interval δt. If δt is short enough the variation of f_{in} during the interval will not matter (it can be made as small as desired by choice of δt) and all the system will care about is its time-integral. In the limit, $\delta t \to dt$, the equivalence of the continuous function and the sequence of impulses is complete. Superposition tells us that the response f_{out} to f_{in} is the sum of all the successive responses to the impulses. Let us define the *impulse response function* $I(z)$ as the form of f_{out} at a time z later than the application of unit impulse, so that the response $\delta f_{out}(t)$ to a single impulse $f_{in}(t')\,\delta t'$ at t' is

$$\delta f_{out}(t) = I(t - t') f_{in}(t')\,\delta t'. \tag{3.2}$$

Time–invariance is implied when $I(z)$ is introduced as a function determined solely by the time interval z between impulse and observation of response, and not by the absolute time at which the impulse was applied. We may now integrate (3.2) to find how f_{out} responds to the whole succession of impulses that make up f_{in}:

$$f_{out}(t) = \int_{-\infty}^{\infty} I(t-t')f_{in}(t')\, dt'. \tag{3.3}$$

This is the most general relation between f_{out} and f_{in} consistent with linearity and time-invariance. The impulse response function may take any form provided $\int_{-\infty}^{\infty} I\, dz$ is finite. If this condition were not satisfied a steady disturbance $f_{in} = $ const, however small, would lead to an unbounded response – which is a way of saying that the system is intrinsically unstable, contrary to hypothesis.

Before proceeding further let us note, without pursuing the matter at length, that an arbitrary choice of $I(z)$ would allow non-vanishing values for negative z; in this case we should find that the response to an impulse would start before the impulse was applied. Leaving aside the philosophical problem of why time appears inexorably unidirectional, a problem to which no satisfactory answer has been given, there is every reason to demand that $I(z)$ vanish for all negative z. If evidence should ever be amassed to allow precognition and related parapsychological phenomena to be mathematically formulated we might find it desirable to modify this imperative demand. It seems safe, however, for the time being to treat the one-sidedness of $I(z)$ as a law of nature having as general a validity as the laws of thermodynamics, to which it is closely linked, since the increase of entropy is meaningful only if time has a unique direction. The effect of this concept of causality in limiting the behaviour of physical systems is summed up in the equations of Kramers and Kronig.[2] This is not, however, a development that needs pursuing here.

Examples of impulse response

Exercise 3.1. A mass m is mounted elastically so that its free oscillations have frequency ω_0 and time constant τ_a for decay of amplitude, assumed slow. Show that the displacement at time $t + z$ resulting from an impulsive force of unit strength, applied at t, takes the form

$$I_x(z) \equiv x(t+z) = (1/m\omega_0)e^{-z/\tau_a} \sin \omega_0 z, \tag{3.4}$$

and that the impulse response of the velocity is

$$I_v(z) \equiv \dot{x}(t+z) = (1/m)[\cos \omega_0 z - (1/\omega_0 \tau_a) \sin \omega_0 z]e^{-z/\tau_a}.$$

$$(3.5)$$

Examples are shown in fig. 3.1(a). If the oscillation is only lightly damped, so that $\omega_0 \tau_a \gg 1$, the second term in I_v is obviously not important.

Exercise 3.2. Now let the damping be such that the oscillations

Fig. 3.1. Impulse response functions for velocity, I_v, and position, I_x, of a harmonic oscillator: (a) underdamped; the tangent to I_v at $z = 0$ crosses the axis at the point $z = \frac{1}{2}\tau_a = \tau_e$, ($b$) overdamped, with $\tau_1 = 2\tau_2$.

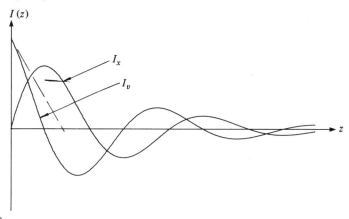

(a)

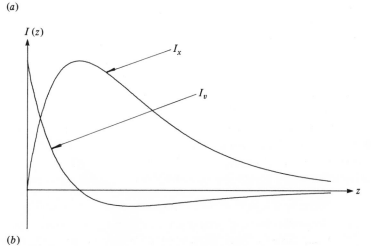

(b)

are overdamped, and the system has two time-constants, τ_1 and τ_2. Show that

$$I_x(z) = \tau_1 \tau_2 (e^{-z/\tau_1} - e^{-z/\tau_2})/m(\tau_1 - \tau_2), \qquad (3.6)$$

and

$$I_v(z) = (\tau_1 e^{-z/\tau_2} - \tau_2 e^{-z/\tau_1})/m(\tau_1 - \tau_2). \qquad (3.7)$$

Examples are shown in fig. 3.1(*b*).

Exercise 3.3. An ideal voltage source (i.e. a generator having negligible internal impedance) is inserted in a series *LCR* circuit. Show that a voltage *V* switched on for a very short time δt sets up a current $V \, \delta t/L$, with no significant charge on the capacitor. Hence show that the behaviour will be analogous to that of the system in Ex. 3.1 and 3.2, with $V \, \delta t$ playing the part of impulsive force, the charge on the capacitor the part of displacement, and the current the part of velocity; furthermore, *m* and *L* are equivalent.

Exercise 3.4. Let the inductance *L* in Ex. 3.3 be reduced in stages to zero. Discuss how the current response changes its form, and show that when $L = 0$,

$$I_i(z) = (1/R)e^{-z/RC}. \qquad (3.8)$$

Exercise 3.5. Repeat the analysis of Ex. 3.3 and 3.4 with an ideal current source (having infinite impedance) connected across the capacitor.

Applications of the impulse response function

Examples to show the use of response functions will make their appearance as this chapter proceeds, but a few preliminary exercises will probably be helpful. The point of the impulse response method is to determine how a linear system responds to some arbitrary time-dependent force, not by trying to solve the equation of motion, including the force, but by breaking down the problem into two parts: first, finding the free motion of the system after a single impulse, and then treating the time-dependent force as a sequence of impulses. The first two exercises will derive well-known results, but the third is not so well known.

Exercise 3.6. Determine the response of the system in Ex. 3.1 to a sinusoidal force of frequency ω, $F(t) = F_0 e^{-i\omega t}$. Note that it is much easier to work with exponentials than with trigonometrical functions, and that the impulse response function in Ex. 3.1 may be written

$$I_x = (i/2m\omega_0)[e^{-i\Omega_1 z} - e^{-i\Omega_2 z}], \qquad (3.9)$$

where

$$\Omega_1 = \omega_0 - i/\tau_a \quad \text{and} \quad \Omega_2 = -\omega_0 - i/\tau_a.$$

Of course, only the real part of the response to $F(t)$ represents the actual behaviour. The result follows immediately from (3.3), which can be rewritten as

$$f_{out}(t) = \int_0^\infty I(z)f(t-z)\,dz. \qquad (3.10)$$

Hence show that

$$x(t) = -F_0 e^{-i\omega t}/m(\Omega_1 - \omega)(\Omega_2 - \omega). \qquad (3.11)$$

8

If the resonance is sharp, i.e. $\omega_0\tau_a \gg 1$, the response has a high peak around $\omega = \omega_0$, where $\Omega_1 - \omega$ becomes very small, while $\Omega_2 - \omega$ is around $-2\omega_0$ and only slowly varying with ω. In these circumstances the more usual Lorentzian approximation is valid:

$$x(t) \sim F_0 e^{-i\omega t}/2m\omega_0(\Omega_1 - \omega). \qquad (3.12)$$

Exercise 3.7. Now repeat the calculation, this time without the assumption that the driving force has been applied for a very long time, but only since $t = 0$. Show that the change in the limits of integration lead to the response (3.12) being supplemented by a transient oscillation at the natural frequency Ω_1, to give

$$x(t) \sim [F_0/2m\omega_0(\Omega_1 - \omega)](e^{-i\omega t} - e^{-i\Omega_1 t}); \quad t > 0.$$

An example is shown in fig. 3.2.[3]

Exercise 3.8. A virtually lossless oscillatory system, having I_x given by (3.9) and with $\omega_0\tau_a \gg 1$, is excited by an oscillatory force whose amplitude F_0 is constant, but whose frequency ω varies

Fig. 3.2. Displacement x of a damped harmonic oscillator driven by a force not at the resonant frequency. The driver is applied at $t = 0$, when $x = 0$, and the initial transient, oscillating at the resonant frequency, beats with the forced oscillation until it dies away.

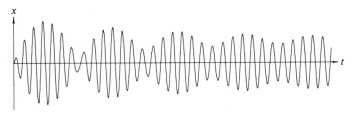

linearly with time from a value much less than ω_0 to one much greater, $\omega = \omega_0 + \beta t$ (this form is chosen so that the moment of resonance, when $\omega = \omega_0$, occurs at time $t = 0$). What is the resulting response of the system to this gliding tone?

Note that the angular frequency is the rate of change of the phase of oscillation, $\omega = \dot{\phi}$, so that $\phi = \omega_0 t + \frac{1}{2}\beta t^2 + \phi_0$. The driving force is then $F_0 e^{-i\phi}$. The analysis is simplified if we neglect losses, so that $\Omega_1 = \omega_0$, and also suppose the second term in (3.9) to be negligible. Show that in these circumstances

$$x(t) = (iF_0/2m\omega_0)e^{-i(\omega_0 t + \phi_0)}S(t) \tag{3.13}$$

where

$$S = \int_{-\infty}^{t} e^{-\frac{1}{2}i\beta t'^2}\, dt'.$$

The meaning of this is most easily seen by the use of Cornu's spiral, fig. 3.3, first introduced to exhibit the solution of straight-edge diffraction patterns (S is Fresnel's integral).[4] Any element $e^{-\frac{1}{2}i\beta t'^2}\, dt'$ is represented on an Argand diagram as a vector of length dt' making an angle $\theta = -\frac{1}{2}\beta t'^2$ with the real axis; the elements are joined end-to-end so that the line joining the end-points represents the (complex) value of $S(t)$. If we measure length s along the curve, starting at the origin, we have $\theta = -\frac{1}{2}\beta s^2$ as the equation defining the curve. The curvature at any point is $|d\theta/ds|$, or βs.

Fig. 3.3. Cornu's spiral; for description see text.

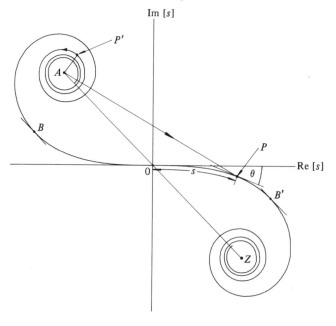

The winding-up of the spiral is the consequence of the curvature being proportional to distance from the origin, measured along the curve. A corresponds to $s = -\infty$, Z to $s = \infty$, and we can locate the upper limit of integration as a point at a distance $s = t$ from the origin, measured along the curve, e.g. P. Then AP represents both the amplitude and phase of the integral. In order to see how $x(t)$ behaves, we must spin the diagram clockwise at angular velocity ω_0, on account of $e^{-i(\omega_0 t + \phi_0)}$, and take the real part. Provided we are indifferent as to the exact phase of the response, we may simply take $|AP|$ as the instantaneous amplitude, and by following its variations as P moves evenly along the spiral discover the time-variation of the amplitude of oscillation; the intensity W, being the square of the amplitude, is shown in fig. 3.4, which you will recognize as the same as the intensity pattern for Fresnel diffraction at a straight edge.

To calibrate the height of this curve it is useful to remember that the length AZ is the same as the distance along the curve between the two points B, B' in fig. 3.3, at which the tangent makes $45°$ to the horizontal; since $\theta = -\tfrac{1}{2}\beta s^2$ these points correspond to $s = \pm(\pi/2\beta)^{\frac{1}{2}}$ and represent a time interval of $(2\pi/\beta)^{\frac{1}{2}}$. Now if the oscillatory force had been applied at frequency ω_0 for this time, the corresponding vector diagram would have been a straight line of the same length. It follows that the ultimate amplitude is the same as if a resonant force F had been applied for time $(2\pi/\beta)^{\frac{1}{2}}$, leading, as you will easily work out, to an intensity $W_\infty = \pi(F_0/m\omega_0)^2/2\beta$. As for the time scale, this can be deduced by observing that the shortest distances from A to points on the spiral occur when θ is approximately $(2n - \tfrac{1}{4})\pi$, or $t = [(2n - \tfrac{1}{4})2\pi/\beta]^{\frac{1}{2}}$.

Finally, let us note that, long after the gliding tone has passed resonance, the residual amplitude A_∞ is close to AZ, which spins at

Fig. 3.4. Time-variation of the response intensity W of an undamped harmonic oscillator to a gliding tone of constant amplitude.

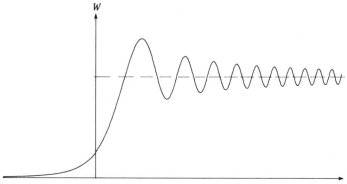

frequency ω_0; the system is thus excited at its own natural frequency and, if no damping occurs, this is what remains. At the beginning, however, the vector AP', representing the amplitude, runs from A to a close point on the spiral, and spins anticlockwise. Since $\theta = -\frac{1}{2}\beta t^2$, the frequency of spinning is $-\beta t$, so that adding the clockwise rotation of the whole diagram at ω_0 we find the frequency of the response to be $\omega_0 + \beta t$, the same as the driver. The system begins by responding to the driver frequency, albeit weakly, but once resonance is passed picks up its own natural motion, leaving the other as a sort of transient modulating the amplitude in the way shown by fig. 3.4.

These three examples have made use of a response function which is the solution of a simple second-order differential equation, and could have been solved directly from this equation without, it might seem, involving a response function. However, the standard procedure for solving this type of differential equation is to discover the impulse-response function, which in mathematical texts is more likely to be called the Green (or Green's) Function.[5] What this chapter has been about, so far, is illustrating in concrete terms a device that was discovered long before there was a clear distinction between science and mathematics. Its use does not depend on knowing the equation of motion for the system – so long as it is linear and I has been determined, whether by mathematics or by experiment, the response to any time-varying force can be calculated. Let us consider just one more example, and an intentionally artificial one, to illustrate the point.

> *Exercise 3.9.* Let $I_x(z) = A$ for $0 < z < \tau$, and 0 otherwise. Show that the oscillation of such a system driven by a force $F \cos \omega t$ will have the form $x = A\tau(\sin\theta/\theta) \cos \omega(t - \frac{1}{2}\tau)$, where $\theta = \frac{1}{2}\omega\tau$. It will not respond at all to a driving force at any frequency which is an integral multiple of $2\pi/\tau$ (cf. the Fraunhofer pattern of a rectangular slit, which is described by the same function).[6]

I have chosen fairly straightforward examples of impulse responses for the sake of illustration, but in practice they may be very complicated, especially if the system is a bell or other structure with a multitude of resonant modes. This only adds complexity, however, without changing the principle.

Compliance, susceptibility, etc.

In these examples we have concentrated on the response of a system to a sinusoidally varying force, partly because the mathematics is easy and partly because of the importance of sinusoidal oscillations.[7] There is no need to work through each case separately from first principles, since a general formula for converting the impulse response into the response to a sinusoid is easily derived. Indeed, it follows almost immediately from (3.3) that if $f_{in}(t) = Ae^{-i\omega t}$, then $\chi(\omega)$, defined as f_{out}/f_{in}, is $\int_{-\infty}^{\infty} I(z)e^{i\omega z}\,dz$. In writing $\pm\infty$ for the limits we recognize for physical reasons that $I(z) = 0$ for $z < 0$, but prefer to keep the entire range of integration to make clear that $\chi(\omega)$ is simply the Fourier transform of $I(z)$. And from the theory of Fourier transforms[8] it follows that if

$$\chi(\omega) = \int_{-\infty}^{\infty} I(z)e^{i\omega z}\,dz, \tag{3.14}$$

then

$$I(z) = \frac{1}{2\pi}\int_{-\infty}^{\infty} \chi(\omega)e^{-i\omega z}\,d\omega.* \tag{3.15}$$

Depending on which branch of physics or engineering is involved, and on the meanings attached to f_{in} and f_{out}, χ takes several names, e.g. see Table 3.1. But the relation of χ to I is the same for all, provided (this may seem a rather obvious proviso) I exists at all and is finite. It is all too easy, however, to assume erroneously that I exists and is well behaved.

Exercise 3.10. When an electric field $\mathscr{E}$ is applied to a metal, the inertia of the electrons prevents them responding immediately.

Table 3.1

f_{in}	f_{out}	name for χ^a
stress	strain	elastic compliance
strain	stress	elastic modulus
electric field	dielectric polarisation	electrical susceptibility
electric field	electric current	admittance, conductivity
electric current	electric field	impedance, resistivity

^{a} Each of these examples of χ has its own symbol, and sometimes several. There are many different elastic moduli and compliances, and electrical susceptibility is χ for unit mass, κ for unit volume. I make no attempt to clarify this confusion. All becomes clear, usually, in the application.

---

* It is more usual to define Fourier transforms so that there is a factor $(1/2\pi)^{\frac{1}{2}}$ before each integral, rather than $1/2\pi$ before one, and unity before the other. If the usual convention is adopted, $\chi(\omega)$ is $(2\pi)^{\frac{1}{2}}$ times the Fourier transform of $I(z)$, and $I(z)$ is $1/(2\pi)^{\frac{1}{2}}$ times the Fourier transform of $\chi(\omega)$. Whichever convention you use, you cannot avoid having to remember the appropriate coefficient.

Instead they accelerate, acquiring velocity at a rate $\dot{v} = e\mathscr{E}/m$, where e is their charge and m their mass. After δt they all have an extra velocity $e\mathscr{E}\,\delta t/m$, and if there are n electrons per unit volume they are at this moment carrying a current density $J = ne^2\mathscr{E}\,\delta t/m$, since $J = ne\bar{v}$. The applied impulse is $\mathscr{E}\,\delta t$, so that the impulse response function for the current starts with an amplitude ne^2/m. After the impulse is over the current is dissipated by collisions, which randomize the directions of motion of the electrons; let us suppose it drops exponentially as $e^{-z/\tau}$ (the model implied in this exercise is very rough in many respects, but it is one that is frequently used – the *free electron model*).[9] Then $I_J(z) = (ne^2/m)e^{-z/\tau}$. Use (3.13) to show that the conductivity $\sigma(\omega) = J/\mathscr{E}$ varies with frequency as

$$\sigma(\omega) = ne^2\tau/m(1 - i\omega\tau).$$

This is well behaved, falling off steadily as ω increases. But now consider the resistivity $\rho(\omega) = 1/\sigma(\omega) = (m/ne^2\tau)(1 - i\omega\tau)$. If you insert this into (3.15) to recover the impulse response function I_ε, being the electric field set up when a current impulse is passed through the metal, you find immediately that the integral is oscillatory and divergent. The mistake lies in believing you can suddenly start a current – the electrons have inertia, and an infinite field is needed for an instantaneous start; I_ε, in fact, does not exist in this case.

It will be noted that $\chi(\omega)$ exists for negative as well as positive frequencies. By replacing ω by $-\omega$ in (3.14) and taking the complex conjugate of both sides, you will see that the real character of I implies $\chi^*(-\omega) = \chi(\omega)$. Hence, from (3.15),

$$I(z) = \frac{1}{2\pi}\int_0^\infty \left[\chi(\omega)e^{i\omega z} + \chi^*(\omega)e^{-i\omega z}\right]\,d\omega = \frac{1}{\pi}\int_0^\infty \mathrm{Re}\,(\chi e^{i\omega z})\,d\omega.$$

We know already that the input function, oscillating at frequency ω, can be represented simply by $f_{\mathrm{in}}e^{-i\omega t}$, which implies $\frac{1}{2}(f_{\mathrm{in}}e^{-i\omega t} + f_{\mathrm{in}}^*e^{i\omega t})$. So here, the output $f_{\mathrm{out}} = \chi f_{\mathrm{in}}$ should be thought of as

$$\tfrac{1}{2}[\chi(\omega)f_{\mathrm{in}}e^{-i\omega t} + \chi(-\omega)f_{\mathrm{in}}^*e^{i\omega t}],$$

or

$$\tfrac{1}{2}[\chi(\omega)f_{\mathrm{in}}e^{-i\omega t} + \chi^*(\omega)f_{\mathrm{in}}^*e^{i\omega t}],$$

which is intrinsically real.

The step-function response

It is sometimes easier, both in experiment and in thinking about a physical problem, to replace the impulse response $I(z)$ by the step-function response $S(z)$. Instead of applying a short impulse one can let f_{in} be $H(z)$, where H is defined as being equal to 0 for $z < 0$ and 1 for $z > 0$. The resulting response to this step-function, measured from $z = 0$, the moment of application, is $S(z)$. Its relation to $I(z)$ is easily found. For if the step occurs at t_0 the force applied between t' and $t' + dt'$ gives rise to a response at time t which is simply $I(t - t') dt'$ and therefore the response due to the steady force is $\int_{t_0}^{t} I(t - t') dt'$. Now write $t - t' = z'$, $t - t_0 = z$; then this integral, which is $S(z)$, has the form

$$S(z) = \int_{0}^{z} I(z') \, dz'. \tag{3.16}$$

Alternatively,

$$I(z) = dS/dz. \tag{3.17}$$

Exercise 3.11. Show that $S(z)$ for the response of current to electric field in the model of Ex. 3.10 is $\sigma_0(1 - e^{-z/\tau})$, where $\sigma_0 = ne^2\tau/m$. The current starts at zero and rises to its limiting value, which determines the d.c. conductivity, as also obtained by putting $\omega = 0$ in (3.15).

The response functions I and S, and the compliance χ, are all equivalent ways of fully specifying the properties of a linear system, and clearly any one can be obtained from any other in a unique way. If one meets χ more frequently than I or S, it is because it is usually easier to measure the frequency variation of compliance, especially at high frequencies, than to study the short time-responses of a system. Modern means of data-collecting, however, which allow impulses to be applied repeatedly until I has been determined to the required accuracy, bring the response functions more into prominence than before. The theoretical examples we shall analyse presently will show, as I have already indicated, how response functions may make it easier to formulate the mathematical problem precisely.

Although we have taken for granted that the response always follows the impulse or the step, the delay is frequently too small to be detected. Suppose, for example, the system is a pure resistor, R, with the applied voltage as f_{in} and the current as f_{out}. Then if τ, as defined in Ex. 3.9, is immeasurably short, the current corresponding to a step-function of voltage is $S(z) = (1/R)H(z)$. Correspondingly the impulse response function $I(z) = (1/R) \delta(z)$, where $\delta(z) = dH/dz$; δ is the Dirac delta-function,[10] which takes the value zero everywhere except $z = 0$, and is infinite at $z = 0$ but has a finite integral, $\int_{-\infty}^{\infty} \delta(z) \, dz = 1$. It therefore describes the gradient of the unit step-function, H. Mathematicians have shown distress at the use of this rather odd function, but physicists have always recognized its practical utility and have seen no problem in using it for the analysis of physical situations. Its pathological properties are rarely, if ever, exhibited in applications. In what follows it is good enough to write $S(z) = 1/R$ rather than $(1/R)H(z)$, since we take for granted that S vanishes for negative z. One must always remember, however, that the derivative of a constant in S appears in I as a delta-function at the origin, not as zero.

Relaxation processes

The simple exponential form $I = Ae^{-z/\tau}$ has already appeared in Ex. 3.4 and 3.10 and is also the correct form for the velocity response of a mass moving slowly in a viscous fluid. Let us note some other examples before proceeding to a general discussion of the compliance coefficient in systems which show this especially simple behaviour.

1. *A liquid polar dielectric, especially a dilute solution of a polar molecule.* A molecule which lacks a centre of symmetry, e.g. CH_3F with three hydrogen atoms and one fluorine arranged tetrahedrally round the carbon, is likely to possess a permanent dipole moment. In gaseous form, or when dissolved in an organic solvent which has no permanent dipole, it is constantly agitated by collisions, and the various molecules produce no net dipole moment per unit volume: $P = 0$. In an electric field, however, they spend longer pointing along, rather than against, $\mathscr{E}$. As a result $P = \kappa\varepsilon_0\mathscr{E}$ in the steady state, κ being the volume susceptibility. All this does not happen instantaneously, and when $\mathscr{E}$ is suddenly applied, P rises gradually to its steady value. Debye,[11] who first studied the matter deeply, suggested that the surrounding molecules behave like a viscous liquid, in which the random motions cause the dipole to undergo Brownian rotations which are slightly weighted in favour of $\mathscr{E}$. He showed that the approach to the steady state was exponential, as indeed it is for almost any model of a dipole in an isotropic environment, not interacting with other dipoles. The response functions follow automatically:

$$S(z) = A + \kappa\varepsilon_0(1 - e^{-z/\tau}), \quad I(z) = A\,\delta(z) + (\kappa\varepsilon_0/\tau)e^{-z/\tau}.$$

The instantaneous response, represented by $A\theta$ or $A\delta$, has been supplied here to indicate that there are sources of polarization other than the reorientation of permanent dipoles. In particular, the electron distribution in each atom or molecule is distorted almost instantaneously on application of an electric field.

2. *Sound waves in a gas whose molecules, such as CO_2, have internal degrees of freedom.*[12] The molecule of CO_2 is linear, and has five degrees of freedom as that term is defined in the context of Boltzmann's equipartition law – three of translation and two of rotation (there is no rotation about the axis of the molecule). If this were all, it would have thermal capacity per mole $C_v = \frac{5}{2}R$, and $C_p = C_v + R = \frac{7}{2}R$; hence $\gamma \equiv C_p/C_v = 1.4$. Measurements show, however, that $\gamma = 1.3$, consistent with $C_v = 3.3R$. The extra $0.8R$ results from partial excitation of the four vibrational modes (two along the axis and two bending modes) which have frequencies considerably greater than $k_B T/h$. In consequence most are unexcited and the mean energy of each mode is much less than the $k_B T$ demanded by Boltzmann's classical law of equipartition. It is also clear that excited molecules only lose their vibrational energy with difficulty. It takes about 10^{-5} s for equilibrium to be established, even though each molecule collides with others about 10^5 times in this interval. If we choose for f_{in} a step-function of temperature, and for f_{out} the extra energy ΔE in the gas at constant volume, we expect to find a step in ΔE of magnitude C_{v0}, the thermal capacity due to the external degrees of freedom (translation and rotation), followed by a gentle increase to a new constant level $C_{v\infty}$, as the vibrational modes become involved. We therefore expect a step-function response something like

$$S(z) = C_{v0} + (C_{v\infty} - C_{v0})(1 - e^{-z/\tau}); \tag{3.18}$$

alternatively,

$$I(z) = C_{v0}\,\delta(z) + (C_{v\infty} - C_{v0})e^{-z/\tau}/\tau. \tag{3.19}$$

Exercise 3.12. Show that $C_v(\omega)$, the thermal capacity for alternating temperature variations at frequency ω, which is the compliance corresponding to S and I in these equations, is

$$C_v(\omega) = C_{v0} + (C_{v\infty} - C_{v0})/(1 - i\omega\tau). \tag{3.20}$$

Hence

$$\gamma(\omega) = \gamma_0 - (\gamma_0 - \gamma_\infty)/(1 - i\omega\tau'), \tag{3.21}$$

where

$$\gamma_0 = 1 + R/C_{v0}, \; \gamma_\infty = 1 + R/C_{v\infty} \quad \text{and} \quad \tau' = C_{v0}\tau/C_{v\infty}.$$

Note that $\begin{cases} \text{when } \omega\tau \ll 1, \; \gamma \sim \gamma_\infty \quad \text{or} \quad 1.3 \text{ if } C_{v\infty} = 3.3R \text{ as in } CO_2. \\ \text{when } \omega\tau \gg 1, \; \gamma \sim \gamma_0 \quad \text{or} \quad 1.4 \text{ if } C_{v0} = 2.5R \text{ as in } CO_2. \end{cases}$

In general, $C_v(\omega)$ and $\gamma(\omega)$ are complex. The meaning of complex $C_{v\infty}$ is that when the temperature is caused to oscillate at frequency ω, the delay in energy exchange between internal and external degrees of freedom causes the energy oscillation to lag behind the temperature. Consider, for instance, what happens as the temperature rises from zero to its peak: the heat input to the external motions follows without delay but the internal 'temperature' lags; even when the temperature is at its peak, at A in fig. 3.5, energy is being transferred from external to internal degrees of freedom, and the source outside the gas must supply this to maintain the temperature of the external degrees. Only a little later, at B, does the transfer from external to internal match the energy loss which the external degrees must suffer to keep in step with the falling source temperature; and this is the moment when the energy input has reached its peak. The variation of ΔE with ΔT, the temperature excess over the ambient temperature, therefore follows an elliptical trajectory, as in fig. 3.5 (you should satisfy yourself that if ΔT and ΔE both vary sinusoidally at the same frequency, but with a phase difference, the resulting trajectory is always an ellipse).

Another way of inducing imbalance between internal and external degrees of freedom is by sudden adiabatic compression of an isolated volume of gas. From a molecular point of view, the rise in temperature simply reflects the increase in mean molecular velocity as some of the molecules bounce off the wall which is moving towards them. At first the extra energy is contained in the translational degrees of freedom, but it is shared almost immediately with the rotational degrees since their energy levels are close together, and every glancing collision can cause exchange of angular momentum. The pressure rises as the volume decreases in accordance with the adiabatic law $p_0 v^\gamma = \text{constant}$, and the appropriate value of γ is γ_0, 1.4 for CO_2. If the volume is now held constant, the external

Fig. 3.5. Relation between injected energy ΔE and excess temperature of source ΔT when ΔT oscillates at such a frequency that the thermal capacity is complex.

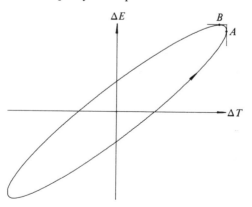

degrees of freedom cool down by giving some of their energy to the internal degrees, and ultimately the pressure is a little less, according to the law $p_\infty v^\gamma = \text{constant}$, with $\gamma = \gamma_\infty$, 1.3 for CO_2. As (3.21) shows, $\gamma(\omega)$ behaves in a very similar way to $C_v(\omega)$, but you should note that the relaxation time τ' is slightly less, by the factor $C_{v0}/C_{v\infty}$.

> *Exercise 3.13.* Represent the internal and external degrees of freedom by two blocks, having temperatures T_i and T_e, and thermal capacities C_i and C_e, which are connected by a weak thermal link, so that the heat current is $\alpha(T_i - T_e)$. Start with $T_i = 0$ and $T_e = T_0$, and hold T_e at this temperature by supplying heat as required. Show that T_i approaches T_e with time constant C_i/α. Now change the conditions so that the whole arrangement is isolated, and both temperatures change to conserve heat. Then you should find the time constant for equilibration of temperatures is reduced to $\alpha(C_i + C_e)/C_i C_e$.
>
> Interpreting this result for the gas, we put $\alpha/C_i = \tau$, $C_e = C_{v0}$, $C_e + C_i = C_{v\infty}$, and hence find that τ', the time constant for temperature equilibrium in the isolated system, is $C_{v0}\tau/C_{v\infty}$ as in (3.21).

Since the pressure takes time, after a sudden adiabatic compression, to reach its final value, we expect, when v oscillates periodically, to find p not following exactly in phase, so that in the range $\omega\tau' \sim 1$ the p–v diagram will look something like fig. 3.5. As a result, the work done in one cycle of volume change does not cancel to zero, but takes the non-vanishing value $\oint p\, dv$, equal to the area of the ellipse in the p–v diagram. This work is irreversibly dissipated as heat, and the effect is revealed in sound 67 propagation, where the relaxation process may result in severe attenuation.

> *Exercise 3.14.* An alternative way of looking at the attenuation of sound is to remember that the velocity of propagation $v_s = (\gamma p/\rho)^{\frac{1}{2}}$, where ρ is the density at pressure p. If γ is complex, following Ex. 3.12, v_s also is complex. Now the equation for plane wave propagation along the x-direction can be written $\psi = \psi_0 e^{i(kx - \omega t)}$, where $v_s = \omega/k$, and ψ represents any scalar quantity (e.g. excess pressure) defining the amplitude of the wave. If k is complex, $k = k' + ik''$, say, $\psi = \psi_0 e^{-k''x} e^{i(k'x - \omega t)}$, and k'' determines the rate of attenuation of the amplitude with distance.
>
> Show that if γ_0 and γ_∞ do not differ greatly, the attenuation is greatest when $\omega\tau'$ is close to unity, and that then the intensity of

the wave, $|\psi|^2$, drops by a factor e in $(\gamma_0 + \gamma_\infty)/2\pi(\gamma_0 - \gamma_\infty)$ wavelengths, or 4.3 wavelengths for CO_2. This is well confirmed by experiment. At 32 kHz, where the attenuation is greatest, it amounts to 117 dB/m.

3. *Anelasticity of solids resulting from dissolved impurities.*[13] The body-centred cubic metals, Fe and Ta, are examples of solids that can carry in solution several per cent of C, O and other atoms. They are dissolved interstitially, as indicated in fig. 3.6. Obviously, so long as the lattice is cubic, all the octahedral sites are equivalent, and must be equally occupied in thermal equilibrium. The solute atoms can hop between sites, albeit infrequently, whenever they happen to acquire from the environment enough energy to take them over the barrier between sites. Any given solute atom may hop on the average only once in a matter of seconds or minutes, but it does allow thermal equilibrium to be established at length. When the crystal is uniaxially compressed, or otherwise distorted, the lattice ceases to be cubic and the sites are no longer equivalent. In equilibrium there will be a higher occupation of the sites of lowest energy. This means that if the distorting stress is held constant after application the process of redistribution will result in a little further change of dimensions. Here again are the conditions for a relaxation process that should result in elastic loss when the crystal is distorted periodically, and the period matches the time constant for redistribution.

Fig. 3.6. Two cubic cells of a body-centred cubic lattice (black circles) with one interstitial atom (stippled) at the centre of an octahedral site. The octahedron is drawn in broken lines. Equivalent positions are found at the centres of all faces and all edges.

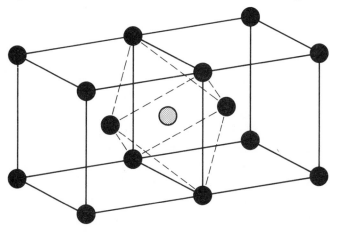

Comparison with experimental results for typical relaxation processes

These examples are typical of relaxation processes which in their simplest form are characterized by an impulse response of the form

$$I(z) = A\,\delta(z) + Be^{-z/\tau}. \tag{3.22}$$

It is by no means uncommon to find processes where the relaxation does not follow a simple exponential. The following analysis leads to a convenient way of presenting experimental data, which will reveal whether the simple form of I is adequate.

The result expressed in (3.20) may be adapted immediately to give the compliance implied by (3.22):

$$\chi(\omega) = A + B/(1 - i\omega\tau). \tag{3.23}$$

If we write χ as $\chi' + i\chi''$, and separate real and imaginary parts,

$$\chi' = A + B/(1 + \omega^2\tau^2) \quad \text{and} \quad \chi'' = \omega\tau B/(1 + \omega^2\tau^2). \tag{3.24}$$

Exercise 3.15. One can represent $\chi(\omega)$ by a curve on the complex plane, each point on the curve corresponding to a different value of ω, as illustrated in fig. 3.7. Show that the curve expressing (3.24) is a semicircle, and prove the properties illustrated, viz. the ends of the diameter are at A and $A + B$, and $\tan\frac{1}{2}\phi = \omega\tau$ at the frequency corresponding to P. 60

You may prefer to approach this in a more general way, which has applications to other problems, by proving that if α, β, γ and δ are complex constants, and x is a real number running between $\pm\infty$, then the points defined by $(\alpha + \beta x)/(\gamma + \delta x)$ describe a circle on the complex plane. The expression (3.23) is a special case, for which inspection shows that the centre of the circle is on the real axis; the ends of the diameter and the position of the centre follow immediately.

Let us compare fig. 3.7 with experiment, remembering that the measurements may need to be manipulated into a suitable form. Thus if the dielectric susceptibility is being studied, the equipment is likely to take the form of an a.c. bridge with a dielectric-filled capacitor in one arm. What is measured is the impedance $Z(\omega)$ of the capacitor, which is $i/\omega\varepsilon C_0$ where ε is the relative permittivity of the dielectric and C_0 the capacitance of the empty capacitor. Hence if $\varepsilon = \varepsilon' + i\varepsilon''$, and $Z = R + iX$,

$$R = \varepsilon''/\omega C_0(\varepsilon'^2 + \varepsilon''^2) \quad \text{and} \quad X = \varepsilon'/\omega C_0(\varepsilon'^2 + \varepsilon''^2).$$

Hence ε' and ε'' are determined; since $\varepsilon = 1 + \kappa$, where κ is the susceptibility per unit volume, it follows that ε and κ have trajectories of the same shape on the complex plane. Fig. 3.8(a) shows a very satisfactory semicircle obtained with heptanol, $C_7H_{15}OH$, though satisfaction is tempered by the realization that in a substance for which ε is over 20 when $\omega = 0$ the dipoles must be in strong mutual interaction; a naive theory based on the assumption of exponential relaxation cannot be taken seriously until the problem has been studied in detail. Our doubts are reinforced by finding that in the transformer oil of fig. 3.8(b), where ε is rather small and the dipoles have a better chance of being independent, the assumption of exponential relaxation is now obviously far from the mark. It is frequently found that the trajectory of ε lies close to a circular arc, though only rarely does the centre lie on the real axis. This curious, and not very satisfactorily explained, behaviour was first noted by K. S. and R. H. Cole, who also pioneered the method of presentation on a complex plane, and have given their name to the plot. The molecules of transformer oil are long, with dipolar groups (whatever they may be) presumably sited somewhere in the chain. When an electric field is switched on, the dipole can twist with its immediate neighbours quite quickly, but it will be some time before the rest of the chain manages to overcome the viscous friction of the surroundings. As it does so, the dipole is able to turn further, and the step response has a long tail. This accounts for the trajectory not being semicircular, but the detailed explanation is not so easily arrived at.

Next let us examine data on the velocity and attenuation of sound in CO_2. The published results present v_s^2 and α_λ, the attenuation per wavelength, as functions of frequency. The former is $\gamma'p/\rho$ and is proportional to γ' if the temperature is kept constant. To obtain γ'' from the attenuation we recall from Ex. 3.14 that the amplitude decays with distance as $e^{-k''x}$ and the wave energy as $e^{-2k''x}$. In one wavelength,

Fig. 3.7. Trajectory of $\chi(\omega)$ on an Argand diagram for a system relaxing according to a single exponential.

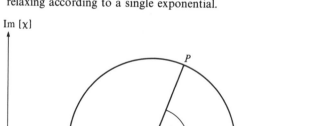

Fig. 3.8. Cole–Cole plots, as in fig. 3.7, of the permittivity ε of (a) heptanol-2 and (b) transformer oil; the points on the curve (b) are frequency in MHz. (c) Analogous plot of v_s^2 in CO_2; frequencies in kHz, v_s^2 in units of 10^3 (m/s)^2. (a) From ref. 3.14, (b) redrawn from ref. 3.15. ((c) From data in ref. 3.16.)

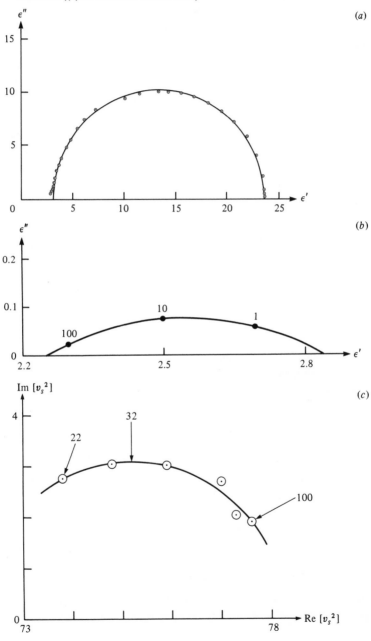

therefore, which is $2\pi/k'$, the energy decays by a factor $e^{-\alpha_\lambda}$, where $\alpha_\lambda = 4\pi k''/k' = 4\pi \tan \theta$, θ being the argument of the complex number k. Since $k = \omega/v = \omega(\rho/\gamma p)^{\frac{1}{2}}$, the argument of γ is -2θ. In the present application α_λ is small enough to make quite negligible the error involved in taking $\tan \theta = \theta$, so that $\gamma''/\gamma' = \alpha_\lambda/2\pi$. Such an approximation would be inadmissable for a diagram like fig. 3.8(a), but here the maximum value of γ''/γ' is only 0.04. Points taken from published data[16] are shown in fig. 3.8(c), and are seen to lie as close to a semicircle as can be expected in view of the experimental errors. The relaxation time can be deduced from the frequency, 32 kHz, at which ε'' reaches its maximum, where $\omega\tau = 1$; hence $\tau = 5 \ \mu s$. This relaxation time is very sensitive to the presence of impurities -0.01% of water vapour halves it. This is probably because H_2O and CO_2 can combine to H_2CO_3 and dissociate again readily. In this process, vibrational energy can be converted to energy of rotation and translation.

The relaxation time in a gas is also pressure-dependent; since the frequency of collisions between molecules is proportional to the pressure, so $\tau \propto 1/p$. It is thus possible to study the relaxation spectrum simply by changing the pressure while keeping ω constant. This technique is not convenient for acoustic studies since the difficulty of launching sound waves increases rapidly with lowering pressure. With dielectric measurements the problem does not arise, especially if the measurements can be carried out at frequency greater than 10^9 Hz, corresponding to a free space wavelength of 30 cm. For the gas can then be contained in a resonant cavity, and the width of the resonant peak is a measure of the cavity losses. The curve in fig. 3.9 was obtained in this way. Since the number of molecules contributing to the loss is proportional to pressure, ε''/p is a measure of the contribution of a given number of molecules to ε'', and it is this that should be compared with χ'' in (3.24).

Fig. 3.9 is an example of another conventional presentation of data on relaxation processes, where χ'' is plotted against $\log_{10} \omega$. When τ is kept constant ω serves to represent the significant variable $\omega\tau$. In the present example ω is constant and $1/p$ is proportional to τ, so that $\log_{10} p = \text{const} - \log_{10} (\omega\tau)$, and the diagram is a mirror image of the more usual representation. This does not matter, since the diagram ideally is symmetrical. If we use natural logarithms and write $x = \ln(\omega\tau)$, it follows from (3.24) that

$$\chi'' = B/(\omega\tau + 1/\omega\tau) = \tfrac{1}{2}B \ \text{sech} \ x; \qquad (3.25)$$

and $\text{sech} \ x$ is a symmetrical function. The width of this curve at half its peak is 2.634 on a natural logarithmic scale, or 1.144 when $\log_{10} (\omega\tau)$ is plotted. The measured curve is slightly wider (1.26), which indicates some small departure from the ideal exponential relaxation.

It is worth pointing out at this point how the logarithmic plot makes the behaviour of χ'', at a superficial glance, resemble a resonance curve. One has only to recollect, however, that the two frequencies at which χ'' is half its peak value, far from being fairly close together as for a resonance curve, are in the ratio 13.93. To make this point even more obvious it is only necessary to draw the curves for χ' and χ'' in (3.24) against $\omega\tau$ rather than $\log_{10}(\omega\tau)$, when any hint of resonant behaviour vanishes altogether (see fig. 3.10).

As a last candidate for detailed discussion, let us look at the measurements of anelasticity (*internal friction*) in tantalum containing about 0.01% of carbon in interstitial solution.[13] Since the interval between successful hoppings of a given carbon atom from one site to a neighbouring site is of the order of many seconds at room temperature, becoming less as the temperature is raised, slow oscillations are needed to reveal the relaxation peak. Kê used a simple torsion wire with an inertia arm to allow free oscillation at about 1 Hz, and determined the imaginary part of the shear modulus from the rate of damping.

Fig. 3.9. Pressure variation, at $-38\,°C$, of the dielectric loss in CH_3F at 1190 MHz.[17]

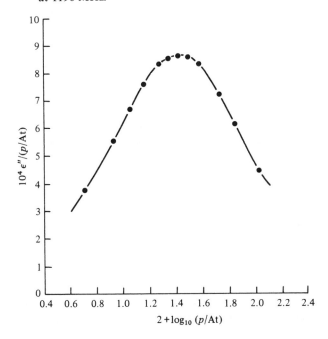

Exercise 3.16. The frequency of a torsion pendulum of given dimensions is proportional to $n^{\frac{1}{2}}$, n $(= n' + in'')$ being the shear modulus at the frequency of oscillation. Show that if the damping is not very rapid, the amplitude falls by a factor e in $n'/\pi n''$ oscillations.

In these experiments it was not convenient to change the frequency, but an equivalent effect may be obtained by changing the temperature. The hopping process by which an atom moves between sites involves crossing a region where its potential energy is high, and which forms a barrier between sites. The atom may be imagined rattling around inside a box and striking one of the walls v times a second, v being approximately a dimension of the box divided by a typical speed ($\sim (k_B T/m)^{\frac{1}{2}}$ for an atom in thermal equilibrium). But if the excess potential energy of the barrier, above that of the box, is W the atom will have sufficient kinetic energy to pass over it for only a small fraction of the time, say $e^{-W/k_B T}$, and only this fraction of collisions will result in a successful hop. Hence $\tau \sim v^{-1} e^{W/k_B T}$, in which we may take v as temperature-independent even though it probably is not. The justification for this cavalier attitude is that if $v \sim 10^{13} \, s^{-1}$, while $\tau \sim 1 \, s$, then $e^{W/k_B T} \sim 10^{13}$ and $W/k_B T \sim 30$. Consequently it is only necessary to raise the temperature by 7 K at room temperature in order to lower $W/k_B T$ to 29.3 and double the hopping rate; this great sensitivity to temperature implies that the relatively slow variation of the pre-exponential factor v^{-1} has very little influence.* Now, as with the last example, we write $x = \ln (\omega \tau) = \text{const} + \ln \omega + W/k_B T$. Then, according to (3.25), if the relaxation process is exponential, a plot of n'' against $1/T$ should yield a symmetrical bell-shaped curve

$$n''(T) = \tfrac{1}{2}B \operatorname{sech} (C + W/k_B T), \quad \text{where } C = \text{const} + \ln \omega.$$

Such curves are shown in fig. 3.11. Their shape accords well with the hypothesis of exponential relaxation with a single relaxation time. The bodily shift when the frequency is changed allows W/k_B to be determined

---

* This argument has much wider implications since it explains why physical and chemical processes taking place on a human time-scale ($\sim$ seconds) are usually very sensitive to temperature ('10-deg rise of temperature doubles the rate' is a good working rule). It requires an activation energy W many times greater than $k_B T$ to slow down the rate from the atomic time-scale to the human. Thus a 2-deg rise in temperature does little to the mean molecular velocities, but has quite an effect on the number of very energetic molecules that can take part in chemical reactions – quite enough to make you feel ill, if it's *your* temperature that rises.

as 12 600 K. This is 33 times 383 K, the temperature at which the measurements show τ to be 1 s, in good accord with our rough estimate.

It is frequently found that the relaxation time is too short for any experiment to be carried out with $\omega\tau \sim 1$. When this is the case (3.24) shows that χ' is essentially constant while $\chi'' \propto \omega\tau$. The phase angle $\chi''/\chi' \propto \omega\tau$ so long as $\chi''/\chi' \ll 1$, and any physical quantity derived from χ shows the same frequency-variation of its phase angle. Thus in sound propagation $k''/k' \propto \omega\tau$, and since $k' = \omega v_s$ it follows that $k'' \propto \omega^2\tau$ – the energy attenuation per unit length, which is $2k''$, is proportional to $\omega^2\tau$. This behaviour is almost universal in liquids at frequencies below 1 GHz, and the magnitude of the attenuation is frequently many times more than can

Fig. 3.10. χ' and χ'', according to (3.24) with $A = 0$ and $B = 1$: (a) against $\omega\tau$, (b) against $\log_{10}(\omega\tau)$.

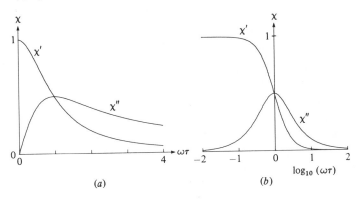

(a)   (b)

Fig. 3.11. Internal friction in tantalum containing interstitial carbon, measured by the decrement of a torsion wire (after Kê[13]). For the left-hand curve $\omega = 7.8 \text{ s}^{-1}$, for the right-hand curve $\omega = 2.0 \text{ s}^{-1}$. The decrement $(1/Q)$ is plotted against the reciprocal of the absolute temperature.

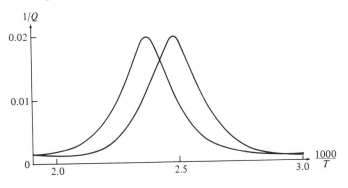

be accounted for by the viscosity and thermal conductivity, both sources of attenuation proportional to ω^2. One must conclude that almost every liquid, after sudden compression, needs time to reorganize its molecular packing. To give one specific example, the charged ions in dissolved salts attract the dipolar water molecules around themselves into a loosely ordered arrangement, which is altered by applying pressure. Sea-water attenuates ultrasonic waves considerably more than does fresh water, and ionic dissipation is the principal reason why echo-sounding cannot be achieved in the sea at frequencies above a few tens of MHz. When the attenuation of sound, or analogous loss mechanisms, varies as ω^2, the magnitude tells one only the product of relaxation strength and relaxation time, $B\tau$ in (3.24). Something else is needed to separate the two; reaching a frequency at which the attenuation begins to flatten off is, of course, one way of doing this.

The approach to equilibrium

In all the models treated in the Exercises, and most of the physical systems discussed, exponential relaxation to the steady state was either the correct description or a close approximation. I shall discuss some other processes in this section to present further examples of exponential approach as well as some clear-cut counter-examples. The exchange of energy between internal and external degrees of freedom in CO_2 was modelled in Ex. 3.13 by two bodies in loose heat contact through a poor thermal conductor. As soon as we extend the model to three bodies, the single exponential pattern of relaxation is lost.

Exercise 3.17. Three bodies, of thermal capacity C_1, C_2 and C_3, at temperatures T_1, T_2 and T_3, are joined by thermal conductances α_{12}, α_{23} and α_{31}. Write down the differential equations for the rate of change of the temperatures, and verify that there are solutions in which all temperatures approach the equilibrium temperature, $T_0 = \sum_i C_i T_i / \sum C_i$, according to the same exponential law e^{-kt}. Do not attempt to evaluate k but show that it is the solution of cubic equation:

$$\begin{vmatrix} C_1 k - \alpha_{12} - \alpha_{31} & \alpha_{12} & \alpha_{31} \\ \alpha_{12} & C_2 k - \alpha_{23} - \alpha_{12} & \alpha_{23} \\ \alpha_{31} & \alpha_{23} & C_3 k - a_{31} - a_{23} \end{vmatrix} = 0.$$

There are three independent solutions, and the response functions must in general contain three exponentials. The process of finding a response function involves a complete solution for the three values of k and, for each value of k, the temperature excess in each body which will relax with this time constant alone. Only then can one construct, say, the step-function response following addition of heat to one body. This is a tedious process and irrelevant to the present discussion, which is solely concerned to illustrate how response functions may arise which are not simple exponentials. Attempts are sometimes made to explain curves like fig. 3.8(b) in terms of a wide spread of exponentials in the response function, but it still remains unclear how this particular distribution of relaxation times is to be related to the molecular structure.

Next let us look at chemical reactions taking place in the gas phase or in dilute solution, or anywhere we are confident that the reacting molecules move independently and meet by chance. In the simplest type, the monomolecular reaction, even meeting is unnecessary – a molecule decomposes spontaneously, perhaps because it happens to have been highly excited by energy communicated from its environment: or, as with a radioactive nucleus, even the environment plays no part, and the emission of a particle occurs at an unpredictable moment. In such reactions the system, molecule or nucleus, does not age – no matter how long it lives before decomposing or emitting a particle, it is the same object up to the very moment of change. The probability that the reaction will occur in the next interval of time, δt, is always the same, whether the system has just been created or has lived much longer than most. If, then, we start with a large population N of identical systems, we may expect that after time t not all will still survive; let us write $NW(t)$ for the survivors. In the next δt a certain fraction, $p\,\delta t$, of these will disappear, leaving $NW(t)$ $(1 - p\,\delta t)$, which of course is $NW(t + \delta t)$. Hence

$$W(t + \delta t) - W(t) = pW(t)\,\delta t,$$

or

$$\dot{W} = -pW,$$

with solution

$$W = e^{-pt}.$$

The coefficient in front of the exponential is unity to ensure $W(0) = 1$. In this case, the relaxation after establishing an unstable population follows a simple exponential law.

When we turn to bimolecular reactions the story is different. In the reaction $A + B \rightleftharpoons C$, the rate at which C is formed depends on the number of chance meetings between A and B, i.e. on the product of their concentrations, while the rate at which C breaks up spontaneously into A and B depends only on the concentration of C. Let us start with x, the concentration of C, zero, and with A and B in the concentrations a and b. Then the reaction is governed by the equation

$$\dot{x} = k[(a-x)(b-x) - \lambda x], \tag{3.26}$$

in which k governs the reaction rate from left to right, and λk the rate from right to left.

> *Exercise 3.18.* Let x_0 be the final concentration of C, when $\dot{x} = 0$, and write $x_0 - x = \xi$. Show that $\dot{\xi} = -k\xi(\mu + \xi)$, where $\mu = a + b + \lambda - 2x_0$. Satisfy yourself that $\mu > 0$, and solve this equation to give
> $$\xi/(\mu + \xi) = (x_0^2/ab)e^{-\mu kt}.$$

Although the relaxation of ξ to zero is not exponential when ξ is comparable with μ, ultimately, when ξ has fallen to a small value, the last stages approximate to $\xi = (\mu x_0^2/ab)e^{-\mu kt}$. This is because one component A (say), has almost gone while B is still present in substantial quantities. Then each molecule of A has a virtually time-independent chance of meeting B, and its final disappearance is like a monomolecular reaction.

Except in special cases, as when $\lambda = 0$ and $a = b$, so that both A and B disappear together, a chemical system after being slightly disturbed from equilibrium will relax exponentially. If it is more complex, requiring more than a single parameter like ξ to describe the departure from equilibrium, each parameter will have its own relaxation time. Normally the behaviour is linear for small disturbances. The case $\lambda = 0$ and $a = b$ in (3.26), however, is governed by the intrinsically non-linear equation $\dot{\xi} = -k\xi^2$, where ξ is the concentration of A and B; $\xi(0) = a$. The solution, $\xi = a/(1 + akt)$, shows ξ falling to zero eventually as $1/t$, not exponentially, since each molecule has to find one of the other kind and both are being depleted together.

As a final example, let us consider a simple quantized system which can occupy either of two energy levels, to see how the relative occupation adjusts itself in accordance with Boltzmann's law. A frequently-met example is an electron in a magnetic field, B; by virtue of its spin and accompanying magnetic moment μ_B, it has two stationary states available to it, of energy $\pm\mu_B B$ (we ignore all other contributions to the energy), corresponding to orientation of the moment parallel or antiparallel to the field. When B is very small, so that the energy difference is negligible, the two states are equally occupied – with an assembly of N electrons, $\frac{1}{2}N$ will be found in each state. When these electrons reside in a solid the thermal motions can induce a change from one state to the other, providing whatever energy may be needed. Clearly when $B = 0$ no energy is needed and the rate of change in each direction must be the same; if n_1 and n_2 are the numbers in the two states

$$\dot{n}_2 = -\dot{n}_1 = k(n_1 - n_2),$$

so that

$$n_1 = \tfrac{1}{2}N + ae^{-2kt}, \quad n_2 = \tfrac{1}{2}N - ae^{-2kt}.$$

When the levels are separated, either by applying B or for any other reason, the rates of transition in the two directions are governed by different coefficients, so that

$$\dot{n}_2 = -\dot{n}_1 = k_1 n_1 - k_2 n_2.$$

We know that in equilibrium, when $k_1 n_1 = k_2 n_2$, n_2/n_1 must equal $e^{-\varepsilon/k_B T}$, where ε is the energy difference between the states, $2\mu_B B$ in the present case. Hence $k_1/k_2 = e^{-\varepsilon/k_B T}$, and it follows that if there is an excess occupation in either state it decays as $e^{-(k_1 + k_2)t}$, always exponentially however far apart the energy levels may be, and however great the displacement from equilibrium.

This argument plays a role in Einstein's famous treatment of black-body radiation,[18] in which he attributed the upward transition rate k_1, to absorption of energy from the radiation field, and the downward transition rate k_2 to spontaneous and to stimulated emission. What is less well known is that on another occasion Einstein[19] suggested that the absorption of sound waves in a chemically reacting gas mixture could be used to determine the relaxation time and hence the rate of reaction. 55

Extended systems

In the foregoing discussion the examples were relatively simple, in that the system was taken as permitting one input and one output function, f_{in} and f_{out}, as it might be the emf across two terminals and the resulting current flow. There is of course no restriction to such an application implied by the theory – in an extended linear system f_{in} and f_{out} may be measured at different places. Electrical networks provide an important and much-studied example, and there are a number of valuable general theorems concerning their linear behaviour of which I shall mention only one. If one arm of the network (labelled i) is broken to allow an ideal voltage source to be inserted, while the current is measured in another (labelled j), then I_{ji} is the time-variation of current following unit impulse of emf. By Fourier transformation we deduce $A_{ji}(\omega)$, the transfer admittance between these points, defining the oscillatory current in the jth arm when unit oscillatory emf is supplied by the source in the ith arm. According to the reciprocity theorem of Rayleigh, generalized by

Carson,[20] almost every linear network, however elaborate, shows the property of *reciprocity*: $A_{ji}(\omega) = A_{ij}(\omega)$: if unit emf in i produces a certain current in j, then unit emf in j produces the same current (and the same phase relationship) in i.* Obviously it follows that $I_{ji} = I_{ij}$ and $S_{ji} = S_{ij}$; and also $Z_{ji} = Z_{ij}$, Z_{ij} being the transfer impedance defined as the emf that appears across an open circuit in the jth arm when unit current is injected into the ith arm.

> *Exercise 3.19.* Verify the reciprocity theorem for the three pairs of arms in the network of fig. 3.12(a); e.g. insert a voltage generator in arm 1 and determine the current in arm 2 to obtain the transfer admittance
>
> $$A_{21} = J_2/V_1 = Z_3/(Z_1 Z_2 + Z_2 Z_3 + Z_3 Z_1).$$
>
> This is clearly symmetrical between 1 and 2, so that $A_{12} = A_{21}$.

Another way of treating this network is to insert voltage generators V_1 and V_2 in two arms, and define the currents i_1 and i_2 as shown in fig. 3.12(b). Then

$$(Z_1 + Z_3)i_1 + Z_3 i_2 - V_1 = 0$$

and

$$Z_3 i_1 + (Z_2 + Z_3)i_2 - V_2 = 0.$$

Hence

$$\frac{i_1}{\begin{vmatrix} Z_3 & V_1 \\ Z_2 + Z_3 & V_2 \end{vmatrix}} = \frac{i_2}{\begin{vmatrix} V_1 & Z_1 + Z_3 \\ V_2 & Z_3 \end{vmatrix}} = \frac{1}{\begin{vmatrix} Z_1 + Z_3 & Z_3 \\ Z_3 & Z_2 + Z_3 \end{vmatrix}}. \tag{3.27}$$

Fig. 3.12. A simple circuit to illustrate reciprocity; (a) and (b) refer to two methods of analysis.

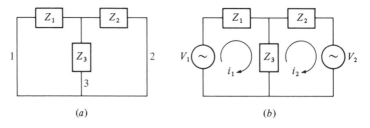

(a)   (b)

* When the network has a magnetic field imposed on it from outside, some of the elements may be affected differently by $+B$ and $-B$; the theorem must then be rephrased as $A_{ji}(\omega; B) = A_{ij}(\omega; -B)$. And similarly, though more rarely, if the network is rotated bodily with angular velocity Ω, $A_{ji}(\omega; \Omega) = A_{ij}(\omega; -\Omega)$.

Now put $V_1 = 0$ to obtain A_{12}, which is i_1/V_2 when $V_1 = 0$; then put $V_2 = 0$ to obtain A_{21}. In both cases

$$A_{21} = A_{12} = Z_3 \left/ \begin{vmatrix} Z_1 + Z_3 & Z_3 \\ Z_3 & Z_2 + Z_3 \end{vmatrix} \right. ,$$

as in Ex. 3.19. In this simple case reciprocity arises because the determinant of the impedances, the third determinant in (3.27), is symmetrical. When the same analysis is applied to a general network the resulting determinant is always symmetrical provided the shared impedances, like Z_3 in this case, are themselves reciprocal. The general proof follows by a straightforward extension of the argument.

This analysis clearly applies to dissipative circuits as well as non-dissipative. In the latter case, where no energy is dissipated as heat, one can prove reciprocity from conservation of energy, and the argument can be extended from networks to continuous structures. An elastic frame of girders, struts, plates, etc., serves to illustrate the principle.

So long as all members behave elastically, the strain energy is uniquely defined by the state of strain at each point, and the total strain energy U is conserved, being altered only by work done on the frame. Now let us apply forces F_1 and F_2 at two points so as to cause, at these points, displacements parallel to the forces and of magnitude x_1 and x_2. An infinitesimal change to F_1 and F_2, giving rise to changes in displacement dx_1 and dx_2, is accompanied by an increase in U equal to the work done by the forces:

$$dU = F_1 \, dx_1 + F_2 \, dx_2. \tag{3.28}$$

Now if these are the only points being disturbed, U will be a smooth single-valued function of x_1 and x_2, since these two displacements determine the deformed equilibrium shape of the whole frame. We can therefore introduce a new function $G \equiv U - F_1 x_1 - F_2 x_2$, and G, like U, is a function of state, being completely defined by the values of x_1 and x_2, or alternatively by F_1 and F_2. Then in an arbitrary infinitesimal change

$$dG = dU - F_1 \, dx_1 - x_1 \, dF_1 - F_2 \, dx_2 - x_2 \, dF_2,$$
$$= -x_1 \, dF_1 - x_2 \, dF_2, \quad \text{from (3.28)}.$$

Hence

$$(\partial G/\partial F_1)_{F_2} = -x_1 \quad \text{and} \quad (\partial G/\partial F_2)_{F_1} = -x_2,$$

so that

$$-\partial^2 G/\partial F_1 \, \partial F_2 = (\partial x_1/\partial F_2)_{F_1} = (\partial x_2/\partial F_1)_{F_2},$$

since the order of differentiation is irrelevant. This result is a statement of Maxwell's reciprocity theorem for elastic structures[21] – if a force $\delta\phi$ applied at point 1 produces displacement δx_2 at point 2, and the same force applied at point 2 produces displacement δx_1 at point 1, then $\delta x_2 = \delta x_1$. For $\delta x_2 = (\partial x_2/\partial F_1)_{F_2}\,\delta\phi$, and $\delta x_1 = (\partial x_1/\partial F_2)_{F_1}\,\delta\phi$, and we have just proved the differential coefficients to be equal. You will know that the same mathematical process applied to the fundamental equation of thermodynamics, $dU = T\,dS - P\,dV$, yields one of Maxwell's relations, $(\partial T/\partial V)_S = -(\partial P/\partial S)_V$, which is another form of reciprocity relation.[22] The others can be derived similarly by introducing H, F and G instead of U. Here again, the justification lies in the existence of U and S as functions of state to which the methods of calculus apply.

It is not so easy to extend this argument to oscillatory forces and displacements, since the frame may then also contain kinetic energy. However, Carson's generalized statement of the reciprocity theorem in electromagnetism gives one confidence that it applies to any linear system in which one can define the analogues of electric field $\mathscr{E}$ and current density $\boldsymbol{J}$, such that $\mathscr{E} \cdot \boldsymbol{J}$ is the power supplied per unit volume. The irreversible thermodynamics of Onsager[23] greatly extends the application (for stationary systems at any rate) by showing how to define generalized forces F_i and corresponding currents (fluxes) J_i in such a way that they obey the reciprocity law: provided the forces and fluxes are linearly related, $J_i = L_{ij}F_j$, then $L_{ji} = L_{ij}$. It is important to follow

Fig. 3.13. (*a*) A stretched string of length 21 units which may be struck at either of the two points 3 or 7 units from one end, and the response measured at the other; (*b*) I_v for both choices in (*a*); (*c*) I_x corresponding to I_v, being $\int I_v\,dt$. The broken line indicates the fundamental period of the string, the time for a pulse to travel the whole length in both directions.

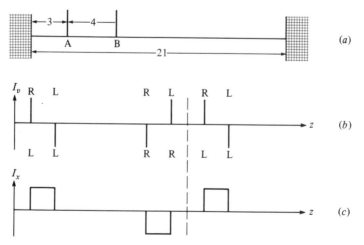

Onsager's prescription for defining F and J, for reciprocity does not apply to arbitrarily selected quantities.

The reciprocity of the impulse response I_{ij} is illustrated by a piano string struck by the hammer. Immediately after the blow we may suppose the string in the neighbourhood, over a length δ, say, to move with velocity v, but the remainder to be at rest. This square pulse of velocity divides into two of height $\frac{1}{2}v$, which move off in opposite directions at the speed of transverse waves. For a good while the pulse propagates more or less unchanged, except that it is reflected with sign reversal at the fixed ends of the string. In fig. 3.13 a specific case is illustrated; the string is chosen to have length 21, is struck at A, 3 units from the end, and the velocity response is recorded at B, 7 units from the same end. The impulse response at B is shown in (b), the letters above each impulse indicating its origin in the pulse that initially moved off to the left (L) or right (R). The letters below show the origin of the same pattern when the hammer strikes at B and the impulse response is recorded at A. Reciprocity is satisfied by the responses being identical in the two cases, but the different manner in which they are produced illustrates that the reciprocity theorem, and the symmetry implied by it, is no superficial matter.

This example is somewhat idealized, as the displacement-time graph of fig. 3.13(c) suggests – the sharp steps, which imply similar steps in the string, are surely rounded off a little. Nevertheless we have here an impulse response resembling the artificial example of Ex. 3.8. Just as there we found the compliance to vanish at certain frequencies, so here the position of the hammer on the string determines the strength with which different harmonics are sounded, and it is possible to arrange that certain harmonics are missing. I leave it as an exercise to apply Fourier transform theory to work out the problem fully.

4

Periodically driven non-linear systems

To exchange the world of linear systems for that of non-linear is to risk the loss of simple guiding principles. No longer can specific solutions be built up by superposition, and the behaviour of a strongly driven system may be entirely different from its behaviour under weak driving. The response may be steady at the driving frequency, or at some other frequency, or it may be unsteady without discernible periodicity. Moreover the response may show sharp discontinuities in form and magnitude as the strength of the drive is changed continuously and slowly. The first approach to rational discussion must be that of the natural historian – collecting typical specimens and hoping that careful examination will yield a helpful classification. Even this must remain incomplete, but the specimens themselves well repay study and the effort to order them is not wasted.

Limit cycles, metastability and hysteresis

As a preliminary let us remind ourselves of the transition from focal stability to focal instability on crossing the horizontal axis of fig. 2.9.   23
The feedback oscillator of fig. 2.13 exemplifies this process. In ideal   28
conditions, as the feedback is raised above the critical value, so that the resistance in the LCR circuit goes negative, the oscillator springs into life, its amplitude increasing exponentially. Thus far linear theory takes us; it says nothing of the ultimate behaviour. We know, however, that the amplifier will begin to saturate and in the end the system will settle down to steady oscillation. On the phase plane of figs. 2.7 and 2.8 the growing   19
spiral will converge onto a closed curve which will be traversed without change. This closed curve is the *limit cycle* for the oscillator. In this particular case, if the circuit resistance is only just overcompensated by feedback the limit cycle approximates closely to harmonic motion. On the other hand, fig. 2.18(*d*) suffices to show that limit cycles need not be   33
sinusoidal. The best-behaved geysers in the Yellowstone Park appear to

be settled in a limit cycle of which most consists of total inactivity, separated by regular spectacular fountains.

Let us examine the limit cycle of the feedback oscillator more closely, assuming harmonic oscillations of charge and current in the resonant circuit. The amplifier, that is, all within the broken lines, may be taken to have an output voltage at Q, $v_Q = \alpha v_P$, when the input voltage at P is v_P, provided $v_P < v_0$; above v_0, v_Q is constant at αv_0. This behaviour is closely followed when a commercial operational amplifier is used. At the oscillation frequency ω it is easy to arrange that ωL_0 is negligible in comparison with R_0, so that the current through L_0 changes in the same way as v_Q, and the emf injected into the resonant circuit is proportional to $\dot{v}_Q$. There will always be some part of the cycle when the current I is near its peak and changes so slowly that the input to the amplifier $v_P = L\dot{I}$, is insufficient to saturate it. Then the injected voltage will be proportional to $\ddot{I}$ and will be greatest when I is greatest. It is during this part of the cycle that energy can be fed into the resonant circuit; for the rest, the amplifier is saturated, $\dot{v}_Q = 0$ and the energy can only be dissipated in the circuit resistance. In the limit cycle the spasmodic energy inputs balance the steady dissipation.

Exercise 4.1. Analyse the circuit in detail to show that

(a) when $v_P < v_0$ the feedback provides negative resistance $\alpha\omega^2 LM/R_0 (= R_1)$ in the resonant circuit;

(b) if the current in the resonant circuit, $I = I_0 \cos \omega t$, the negative resistance only operates while $|\sin \omega t| < v_0/\omega LI_0$, but the circuit resistance R operates at all times;

(c) the voltage amplitude of the limit cycle, measured at P, is v_0/x, where

$$\sin^{-1} x + x(1-x^2)^{\frac{1}{2}} = \pi R_0/2R_c, \tag{4.1}$$

and R_c is the value of R_0 at which oscillation begins, i.e. $\alpha\omega^2 LM/R$.

It is not too difficult to set up the circuit to test this result. An air-cored inductor of about 40 mH, with 50 turns wound round the outside to provide M, serves well. The points in fig. 4.1 were taken in much less than an hour, using a digital a.c. voltmeter to measure the amplitude (up to 14 V). R_c is determined directly, and indeed so may v_0 be determined, but I allowed myself a little adjustment to get the best fit. The required value agreed adequately with the direct measurement. It will be seen that over most of the range the amplitude V varies as $1/R_0$; when x is small, the left-hand side of (4.1) is close to $2x$, so that $V/v_0 \sim 4R_c/\pi R_0$.

In this case, and whenever the amplifier characteristic is uninflected so that the effective amplification decreases monotonically with input amplitude, the circuit enters and leaves oscillation reversibly, without hysteresis, as the negative resistance is made greater than, or less than, R. I measured the critical value of R_0 first increasing and then decreasing, and found agreement to better than 1 in 1000. An inflected amplifier characteristic, like that in fig. 4.2(a), would have altered this significantly. It is obvious that if v represents the amplitude of the sinusoidal oscillations of the resonant circuit, the negative resistance R_1, averaged over the complete cycle, will behave somewhat as in fig. 4.2(b). The vertical scale of this curve varies inversely as the resistance R_0 in the feedback circuit, and $R_0 R_1$ is therefore a unique curve. On the same diagram let us plot $R_0 R$, a horizontal line representing the fixed resonant circuit resistance, but now at a level fixed by R_0. As the feedback is increased, by dropping the value of R_0, the line which starts at (1), say, falls to give the first intersection P_1 with the curve of $R_0 R_1$ (line 2). Spontaneous oscillation will not occur since, unless an exceptionally large fluctuation raises the amplitude to v_1, $R > R_1$ and any incipient oscillation is damped. Only at (3) do negative and positive resistances cancel for zero amplitude,

Fig. 4.1. Amplitude of oscillation in feedback oscillator ($\propto 1/x$) against strength of feedback (R_c/R_0): logarithmic plot.

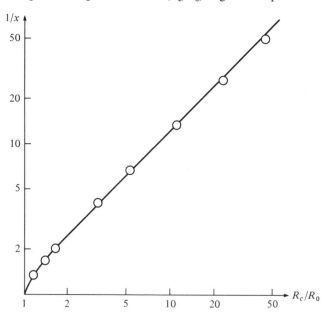

allowing spontaneous oscillation. At this point the amplitude rises sharply to P_2, for anywhere inside the curve for $R_0 R_1$ the negative resistance is more than enough to overwhelm R. Thus only the right-hand part of the curve, beyond P_1, is realizable in the steady state, and it can be used to construct fig. 4.2(c) which shows how the amplitude of oscillation varies with $1/R_0$. It should be noted that as R_0 is increased to reduce the strength of feedback, the already-oscillating circuit continues to oscillate until its amplitude falls to v_1; at this moment the intersection in (b) has risen to P_1. The imminent collapse of the oscillation is signalled by the increased slope of the curve for v against $1/R_0$. It becomes infinite before the curve has fallen to zero, and there is no possibility of catching the oscillation in a

Fig. 4.2. Hysteresis in a feedback oscillator: (a) assumed characteristic of amplifier; (b) curve showing how the negative resistance varies with the amplitude of oscillation. See text for further explanation; (c) the hysteresis curve followed as the feedback is changed.

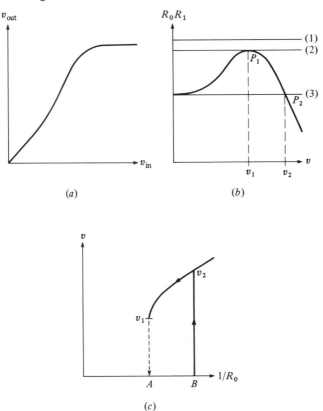

steady state with any amplitude less than v_1. On the way up to the point of spontaneous oscillation there is no such precursor to warn of an imminent instability. Here, however, as in the transition from focal stability to focal instability, the warning is to be found in the time-constant τ_a for decay of small oscillations, which rises to infinity at the critical point.*

What we have described is a typical example of *hysteresis*, arising from a particular type of non-linearity. Between A and B in curve (*c*) the non-oscillating system is said to be *metastable*, like, for example, a rod balancing on its squared-off end; it is stable against small displacements, but large displacements reveal a preferred state of absolute stability – the rod lying flat, the circuit oscillating in its limit cycle.

Any maintained oscillatory system is liable to be hysteretic around the onset of oscillation if the frictional forces vary more slowly than linearly with velocity (or current). Such behaviour is common enough in mechanical systems. The clock escapement is so extreme an example as almost to pass unrecognized – not only is there sliding friction, but the feedback mechanism is not even actuated until a certain amplitude is reached.[1] In electrical circuits the effect is rarer, though it may readily be synthesized by a slight extension of the feedback circuit, fig. 2.13. All that is needed is a second amplifier supplying a feedback circuit and mutual inductance in parallel with the first; this time, however, the feedback must provide extra positive resistance and have a larger amplification, so that it is the first to saturate. There is then a regime of low voltage where the effective circuit resistance is large, perhaps too large for oscillation to be maintained: an intermediate regime, after the extra resistance has disappeared by reason of saturation, where the original feedback may be more than enough to give net negative resistance: and a high-voltage regime where both amplifiers are saturated and positive resistance is restored. Fig. 4.3 shows the hysteretic response as the magnitude of the negative resistance is altered. Unfortunately this circuit does not show, like fig. 4.2(*c*), the characteristic vertical slope and discontinuity as $1/R_0$ is reduced, but only a virtually unheralded discontinuity. This is because the

---

* The collapse of the oscillation, when the feedback is reduced to a point beyond which no oscillatory solution exists, is an example of a *limit point instability*. The spontaneous oscillation that occurs as the feedback is increased is different in kind. For the circuit can remain unexcited in principle, provided it is not disturbed; the unexcited state is analogous to unstable equilibrium. The transition from a stable to an unstable unexcited state is an example of an *unstable symmetric transition*. Both these transitions are discussed at length in chapter 5.

sharp knee in the amplifier characteristic, where saturation sets in, causes the feedback resistance to have a discontinuity in its derivative. Fig. 4.2 was based on smooth characteristics which are on the whole more likely in physical examples.

Mode jumping

The phenomenon just described is representative of a common occurrence in non-linear systems, by no means restricted to those in which the resistance or friction term provides the non-linearity. Sudden hysteretic discontinuities of response occur when the natural frequency of a passive vibrator varies with the amplitude of oscillation. Such a system, driven by an oscillatory force, may respond vigorously at first, if the force is tuned to the low-amplitude natural frequency, but fail to maintain growth as the larger amplitude brings detuning with it. Alternatively, the process may begin with the driving force somewhat detuned, but if it is strong enough it may drive the system to an amplitude of response at which it is in resonance. Naturally there will then be a sudden improvement in the response, which may well show a discontinuity. To carry the argument further we need a diagram.

Let us suppose that the system behaves like a particle of mass m vibrating in a one-dimensional potential well which, instead of being quadratic, $V = \frac{1}{2}\mu x^2$, rises more sharply at larger x: perhaps $V = \frac{1}{2}\mu x^2 +$

Fig. 4.3. Measured hysteresis curve, to compare with fig. 4.2(c).

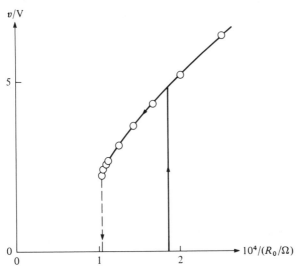

λx^4. Then, as the amplitude is raised, the natural frequency will increase above its low amplitude value:

$$\omega_0 = \Omega + \beta A^2 = \Omega + \beta I, \tag{4.2}$$

where $\Omega = (\mu/m)^{\frac{1}{2}}$, β is a constant and A is the amplitude; A^2 is written as I, the intensity of the response. We do not need to know how β is related to the form of V; it is enough to indicate the line for $\omega_0(I)$ on the diagram of fig. 4.4. We now make an important assumption: if ω_0 is not strongly dependent on I, and the potential well is very nearly quadratic, we take the form of the vibration to approximate so closely to sinusoidal that the real potential may be replaced by an effective quadratic potential, such as would give rise to vibration at frequency ω_0. At every value of I the potential must be appropriately adjusted. The steady-state response to a sinusoidal driving force can then be expected to match that of the equivalent harmonic oscillator. If the loss is low, so that a truly harmonic system would show Lorentzian response to a driving force of constant strength and variable frequency, as in (1.10), then (4.2) shows that

$$I = I_0/[(\omega - \Omega - \beta I)^2 + \lambda^2]. \tag{4.3}$$

This is a cubic equation in I whose solution is most easily computed by finding ω as a function of I. The responses for various amplitudes of driving force are shown on fig. 4.4. It will be seen that weak forces bend the symmetrical Lorentzian curve over to the right, since the line $\omega_0(I)$ bisects

Fig. 4.4. Theoretical resonance curves for a non-linear vibrator, showing how the square of the amplitude of vibration (the intensity, I) varies with the frequency of a constant amplitude driving force. The three curves are for different strengths of driver.

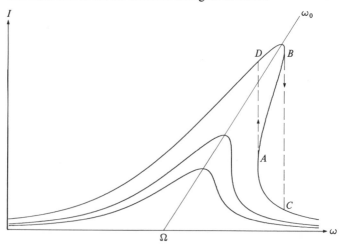

points of equal I. Above a certain critical strength of drive, represented by the middle curve, the solution for $I(\omega)$ has a region of many-valuedness, between the broken lines for the uppermost curve.

> *Exercise 4.2.* Show that (4.3) may be written as $X = CY \pm (1/Y - 1)^{\frac{1}{2}}$, where $X = (\omega - \Omega)/\lambda$, $Y = \lambda^2 I/I_0$ and $C = \beta I_0/\lambda^3$. The two signs refer to points on either side of the line $\omega_0(I)$. Note that when C is chosen to give the critical response curve (the middle curve in fig. 4.4) there is a vertical inflexion on the curve, where both dX/dY and d^2X/dY^2 vanish. Hence show that the critical value of C is $8/3\sqrt{3}$. You may recognize that the argument exactly parallels that used conventionally to find the critical point in van der Waals' equation, which is a horizontal point of inflexion of the P–V curve.[2]

It is easy to see that the stretch between A and B in fig. 4.4 is unrealizable. Suppose the oscillation was poised at a point on that stretch. If it happened to suffer a small increase of amplitude, it would find itself inside the Lorentzian curve, where the amplitude is less than the driving force is able to maintain at that frequency. Consequently the amplitude would rise until it hit the upper line. Similarly, if the amplitude fell a little, it would find it was too high to be maintained, being outside the Lorentzian, and would drop to the lower line. The behaviour is now clear. As ω is increased from below Ω, the resonance curve is followed as far as B, at which point the amplitude collapses to C, and from then onwards the curve is single-valued and is followed automatically. On the way back the amplitude rises to a vertical tangent at A, and jumps to D, thenceforth following the single-valued curve back to low frequencies. It should be noted that we have discussed the stationary solutions, and have said nothing about transient effects. The jumps from B to C and from A to D are only discontinuous in the sense that the steady-state behaviour is discontinuous. When the experiment is done the jumps tend to overshoot and are followed by transient pulsations of amplitude which gradually die away. It is worth noting that the characteristic foreshadowing of a hysteretic discontinuity occurs at both jumps, the slope of the curve becoming infinite before the jump. Both jumps are limit point instabilities.

The apparatus shown in fig. 4.5 may be used to demonstrate the above effects. The strip of melinex (mylar) $\frac{1}{2}''$ wide and $0.004''$ thick, stretched on an aluminium frame, is highly non-linear in torsion; when the magnets fixed to the centre are twisted about a vertical axis, the restoring torque rises more rapidly than linearly. If melinex is not available, two pieces of sticky plastic tape (e.g. Sellotape in UK) stuck together by their sticky faces, are an adequate substitute. The coil shown, which is one that happened to be available, is larger and certainly better made than is necessary to provide the oscillatory horizontal magnetic field that excites the magnets. A simple signal generator and audio amplifier gave more than enough field for the purpose. Fig. 4.6 is a typical response curve for constant driving field and variable frequency. The rough measurements were made to allow a fair curve to be drawn; in the laboratory the jumps and hysteresis are so dramatic that qualitative observation suffices to convince.[3]

There are many examples of mode-jumping in electronic circuits which can be understood in principle without difficulty, but which are not readily amenable to quantitative treatment. The applied mathematician would say the same about the non-linear oscillator, which we have dealt

Fig. 4.5. Experimental arrangement to demonstrate mode-jumping. The non-linear vibrator is the bar magnet M mounted on a plastic strip S kept in tension by an aluminium frame F. The magnet is caused to vibrate by passing an alternating current from a signal generator through the solenoid B, and the amplitude can be roughly read off the protractor.

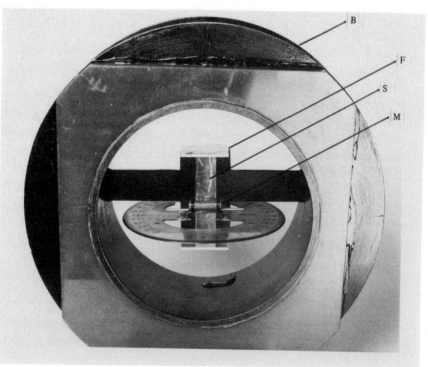

with summarily by assuming its motion to be near enough sinusoidal for standard resonance theory to apply. To justify this assumption, or rather to determine how great is the error thus introduced, is not trivial and is fully discussed in standard texts on the mathematical aspects of non-linear systems, to which (without enthusiasm) I provide references.[4] There it will be found that (4.3) is indeed a sound first approximation.

In the same intuitive spirit let us look at some other examples of mode-jumping. Suppose that the feedback oscillator of fig. 2.13 is used to excite a second resonant *LC* circuit, without any precautions to prevent the response reacting back on the oscillator. The simplest arrangement is to couple the second *LC* circuit to the first by a mutual inductance. Then, in effect, the two coupled circuits can be thought of as a single system having two modes of oscillation, either of which may be maintained by the feedback circuit.[5] Let us first consider the behaviour before feedback is applied. It is perhaps easier to visualize what happens by thinking of an equivalent mechanical model, in which the circuits are replaced by pendulums, coupled as in fig. 4.7(*a*) by being hung on a slack string. If you have never played with this system it is well worth doing so. All you need is some string draped between two chairs and a couple of bottles or a pair of shoes to act as bobs. When the pendulums have the same mass and length, there are two normal modes which preserve their pattern until the oscillations die out: in one the pendulums swing in phase, with the same amplitude, and rather more slowly than in the other mode, where they

Fig. 4.6. Hysteretic response curve measured with the set-up of fig. 4.5. The amplitude θ is shown as a function of the driving frequency ν.

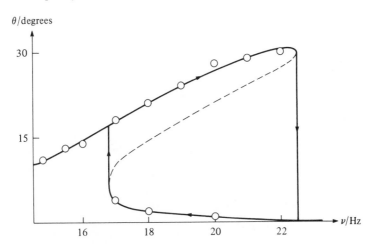

swing with the same amplitude in antiphase. The frequencies Ω of the symmetric and antisymmetric modes are shown as s and a in fig. 4.7(b). Now keep pendulum 1 unchanged and alter the length of pendulum 2; you will find that when pendulum 2 is short, and has a substantially higher natural frequency than pendulum 1, the two modes take the form of one pendulum swinging vigorously, and exciting a weak swing in the other. Thus at H_2 pendulum 2 swings at something close to its natural frequency, and because pendulum 1 is being forced to swing at a higher-than-natural frequency for it, it swings weakly in antiphase. By contrast, the mode at H_1 has pendulum 1 dominant and pendulum 2 a weak in-phase parasite. Similarly when pendulum 2 is lengthened, it is at L_1 that pendulum 1 dominates, and at L_2 that pendulum 2 dominates.

> *Exercise 4.3.* Show that the frequency of the mode labelled s is that of a pendulum whose point of support is on the line PQ vertically above R; that the effective point of support for a is vertically above R on the line PT; and that the limiting frequencies for H_1 and L_1 are for a point of support on PS, vertically above R.

After this, it should now be plausible that when an escapement is attached to maintain pendulum 1, there is a limit to how far along the branches H_2 and L_2 can be maintained; for in these modes all the action takes place where the escapement is not operative. Thus we need not be surprised to find that when pendulum 2 is tuned to a higher frequency, the

Fig. 4.7. (a) Two identical pendulums hang from a slack string; (b) effect of changing the length of pendulum 2 on the normal modes; Ω is the frequency of a normal mode, ω_1 the (constant) frequency of pendulum 1, and ω_2 the (variable) frequency of pendulum 2.

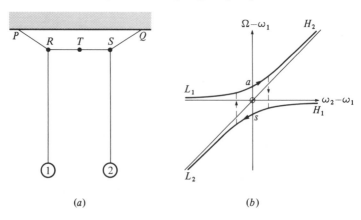

(a)   (b)

system vibrates at a frequency less than ω_1, as at H_1; and conversely, when pendulum 2 is tuned low, the system vibrates at a rather higher frequency, as at L_1. These two points belong to different branches, and if the coupling is strong, so that the branches are well separated, the operating frequency does not proceed smoothly from one to the other as the length of pendulum 2 is altered, but discontinuously, with hysteresis, as sketched in the figure.

Although it is perhaps easier to visualize the mechanical model, the electrical circuit is undoubtedly easier to realize. Fig. 4.8 shows the curves obtained when the natural frequency of the second circuit is varied continuously by changing its capacitance, C_2. The behaviour follows along the expected lines, with discontinuities and hysteresis, when the circuits have high Q, so that the splitting is large compared with the natural widths of the resonance. There is a close parallel between this and the behaviour of the non-linear oscillator in fig. 4.4, in the sense that both show a continuous response when Q is low, but become discontinuous and hysteretic at high Q.

Fig. 4.8. Mode-jumping in a feedback oscillator whose *LC* circuit is inductively coupled to a passive *LC* circuit. The natural frequency of the latter is changed by means of its capacitor C_2. The coupling is strongest for curve *a*, which is highly hysteretic, and weakest for curve *c* which shows no critical behaviour.[6]

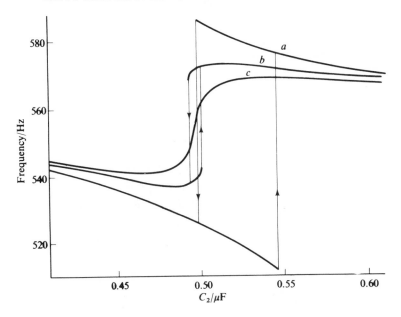

If you can find a metal tube about 3 cm diameter and 60–100 cm long, you should be able to observe the phenomenon without further trouble. Simply put one end in your mouth and, with your teeth wide but your lips gripping the tube, hum a sliding note through the whole gamut of your voice. You should notice at least one point where the note jumps, and you will feel, quite as well as hearing, the instability.[7] Whether you will notice hysteresis between a rising and falling note will depend on how well your vocal chords and resonant cavities are coupled to the resonances of the tube.

Something similar, and troublesome, was found during the early development of radar, arising from coupling a powerful oscillator (e.g. a pulsed magnetron) into a waveguide feeding the aerial. Although reflections in the waveguide and at its exit to the aerial were kept as low as possible, there was always a weak reflected wave back at the oscillator; and the phase of this reflection varied rapidly with frequency if the waveguide was many wavelengths long. To see how this might cause trouble, consider a model experiment in which the oscillator is the same feedback oscillator as before, and let the output be taken by a mutual inductance to a long cable with a weak reflection at the end. Without bothering about details of transmission line theory,[8] let us simply note that the transmission line will couple into the LC circuit of the oscillator an additional impedance varying periodically with frequency. The variations of resistance will result in slight changes of the level of oscillation, which we ignore. More important are the variations of reactance.

> *Exercise 4.4.* (a) If the transmission line provides in the resonant circuit an extra, frequency-dependent reactance $j\Delta \cos \beta\omega$, show that the equation determining the resonant frequency is
>
> $$\omega L - \frac{1}{\omega C} = -\Delta \cos \beta\omega.$$
>
> (b) Show that for a transmission line of length l, $\beta = 2l/v$, where v is the velocity of waves on the line. This choice of β ensures that a change of frequency sufficient to add one more wavelength to the path along the line and back again, is also such as to change $\beta\omega$ by 2π.
>
> (c) If the line is many wavelengths long, and the reflection at the end weak enough so that ω is not pulled far from $\omega_0 = (LC)^{-\frac{1}{2}}$, write $\omega/\omega_0 - 1 = x$ and show that, approximately,
>
> $$2\omega_0 L x = -\Delta \cos (\beta\omega_0 x + \theta), \quad \text{where } \theta = \beta\omega_0.$$

(d) Find the solution to this equation by drawing both sides as functions of x and locating the intersections. Show that one solution is guaranteed only if $\Delta/\omega_0 L < 1/2\pi N$, where N is the length of the transmission line in wavelengths.

(e) Finally, use this diagram to discuss how the resonant frequency changes when L is slowly varied. If the condition (d) is not satisfied, the resonant frequency has smooth changes alternating with jumps.

The parallel between this behaviour and that of fig. 4.6 is obvious, with the transmission line playing the role of a resonator coupled to the maintained oscillator. In a microwave radar transmitter the waveguide output is rather tightly coupled to the oscillator, to obtain maximum output, and the waveguide itself may be well over 100 wavelengths long if λ is 1.5 cm. For single mode operation this demands that $\Delta/\omega_0 L < 1/600$, and the need to eliminate reflections in the line to bring Δ down to so low a value creates serious design problems. Fortunately there now exist non-reciprocal circuit elements, called *isolators*,[9] which allow free passage of power from transmitter to aerial, but absorb reflected waves.

This discussion is readily generalized to real microwave transmitters such as the resonant-cavity magnetron. By empirical measurements one can construct for a given transmitter contours of constant output frequency on an impedance plane, or on a chart representing the amplitude and phase of the reflected wave. This chart, which describes the properties of the transmitter alone, is called a Rieke diagram.[10] In any given application, the amplitude and phase of reflection of the transmission line, as functions of frequency, are represented as a trajectory on this same diagram; any point where the trajectory intersects a contour at the same frequency represents a possible mode of the combined transmitter and waveguide. The system can then be designed to avoid ambiguities such as may cause the transmitter to switch modes at the smallest disturbance.

Finally, in this catalogue of mode-jumping, reference may be made to the phenomenon of entrainment, or phase-locking, first noted by Christiaan Huygens, the inventor of the pendulum clock. He found that two pendulums, each maintained by its escapement, when mounted on the same wall (which must have been less substantial than one would infer from Dutch interior paintings of the period) pulled into synchronism so that thereafter they kept a constant phase-relationship.[11] A simpler demonstration involves coupling a signal generator output inductively into a maintained LC oscillator.[12] It will be found that when there is a considerable frequency difference between the two, the oscillator continues at its own frequency, with the amplitude modulated into a beat pattern at the difference frequency. So far there is little difference from the behaviour of a freely swinging pendulum disturbed by a non-resonant force. The difference begins to be apparent when the signal generator is tuned towards the oscillator frequency. The beat period increases until, quite suddenly, the oscillator locks into synchronism; even though it is now oscillating at other than its own natural frequency, it has been trapped by the injected signal and the beat pattern vanishes. When the signal generator is detuned again, the oscillator suddenly snaps out and reverts to its natural frequency, beating with the injected signal; the behaviour is usually hysteretic.

Highly non-linear systems; introduction to the Poincaré map

We turn now to systems whose natural frequency is strongly dependent on their energy, and ask what happens when they are driven by a constant-frequency force of variable amplitude. Examples which have been studied in some detail include the pendulum driven so strongly that it may be forced right over the top.[13] The equation of motion is

$$ml^2\ddot{\theta} + f(\dot{\theta}) + mgl \sin \theta = G(t),$$

in which $f(\dot{\theta})$ is a velocity-dependent frictional couple and $G(t)$ is the applied couple. This equation is the same as would be obeyed by a particle moving in one dimension in a potential proportional to sin kx. Again, in what has come to be known as the Duffing oscillator[14] the potential $ax^2 - bx^4$ has a quadratic minimum at the origin, reaches a maximum of $a^2/4b$ when $x^2 = a/2b$, and then falls indefinitely. The frequency drops to zero when the energy is $a^2/4b$, and above that the particle is not contained, just as the pendulum with energy greater than $2mgl$ turns continuously (though far from evenly) instead of undergoing oscillations. In both cases, however, a strong oscillatory driving force may constrain the motion so that the system goes over the top, but is brought back before it can escape.

119

Obviously under these conditions it is quite wrong to suppose the motion to be almost sinusoidal, as we have imagined so far. Unfortunately, whether the potential is sin kx or $ax^2 - bx^4$ – or, for that matter, practically everything except x^2 – the equation of motion of the driven system has no analytical solution, and we must be content with approximate methods, which are complicated and not very rewarding for our purposes, or computer solutions which are slow except on a large computer. On the other hand, there are special highly non-linear systems which can be synthesized from sections, each of which is analytically tractable, and for which the complete solution can be specified by matching boundary conditions between the sections. One example is a ball which is kept continuously bouncing off a vertically oscillating base-plate.[15] Most of the time it is in free fall, and the evolution of its motion is found by determining how fast the base-plate moves at the moment of impact, and hence how fast the ball is thrown up again for its next flight.

Here we shall consider another example, which I shall call an *impact oscillator*, and is simple enough to be handled by a small programmable calculator. It exhibits all the properties shown by the more conventional models that demand heavy computation. Herein lies the justification for devoting considerable space to a problem that, though realistic physically, can hardly be regarded as important in itself. Unfortunately the genuinely important examples are still beyond the reach of exact analysis. We therefore turn to fig. 4.9, showing an elastic ball, with coefficient of restitution ε, bouncing between two walls; these walls are fixed to a rigid frame which is caused to vibrate to and fro so as to maintain the motion of the ball. Before tackling the analysis, however, there are several important preliminaries, the origin of which is illustrated by this model. It may be that the ball is captured in a stationary pattern in which it bounces in synchronism with the vibration of the walls, reversing its velocity at each bounce. In this case the trajectory in phase space is like fig. 4.10(a), with the ball striking each wall when it is beyond its midpoint, but moving inwards.

Fig. 4.9. The impact oscillator. The two side walls are fixed $2l$ apart and vibrate as $a \sin \omega t$. The ball is constrained to move along a horizontal line, bouncing against the walls.

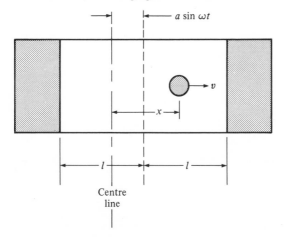

Centre
line

On the other hand, if the ball is not so captured the trajectory is much more complex, as in fig. 4.10(*b*). If we were considering a less extreme example, such as the Duffing oscillator, the trajectories would be smooth. There is now no bar to their intersecting one another, as there is for a freely running system; for the behaviour is no longer dictated solely by the point on the phase plane, but also by the phase of the driver, which may be different on successive returns to a given point.

Suppose we record the co-ordinates (x, v) reached by the system at a chosen moment in the driver cycle. This is enough to define all subsequent motion since the time of arrival will be known at each point reached by the representative point in phase space, and hence the driving force. A full account of the behaviour does not demand drawing the whole pattern as in fig. 4.10; it is enough to follow what is going on if we record successive values (x_n, v_n) once per cycle, every time the phase of the driver reaches a predetermined value. And, provided the amplitude of the driver is kept constant, there is a unique relation between subsequent points:

$$(x_{n+1}, v_{n+1}) = F(x_n, v_n). \tag{4.4}$$

This equation, whose form must be found by integrating the equations of motion over one cycle, starting from an arbitrary point (x, v), defines a one-to-one relationship, called a *mapping*, by which points (x_n, v_n) are transformed into (x_{n+1}, v_{n+1}). What I am now describing is a *Poincaré map*. Since the equations of motion are not usually soluble in closed form, the function F has usually no analytical form, but the model of fig. 4.9 is an exception, and in due course we shall determine F in this case.

Fig. 4.10. x–v phase plane trajectory of the ball in (*a*) a stationary pattern and (*b*) a non-stationary pattern. A and B mark the mean position of the walls.

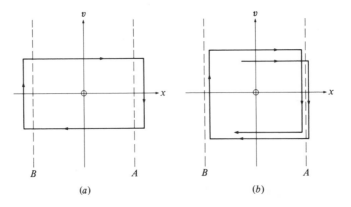

(*a*)   (*b*)

Let us not run ahead too fast, however. There is a special case of (4.4) that must be considered first, arising when the state (x, v) is a stationary state such that $(x, v) = F(x, v)$; such a state is pictured in fig. 4.10(a). What determines whether it is stable against small perturbations? We tackled a problem like this in chapter 2, and now must extend the analysis to cover discrete sampling. Only when we have dealt with stationary states and proceed to non-stationary solutions do we need to understand the extraordinary possibilities inherent in (4.4). And since the idea of mapping is likely to be novel I shall, as a preliminary, devote some space to the one-dimensional mapping problem, $x_{n+1} = f(x_n)$, to reveal the subtleties lurking in so innocent-looking an expression. But first the question of stability.

Stability rules for discrete sampling

When the system is caught in a stationary trajectory, which it repeats at every sampling, $x_{n+1} = x_n = x_0$ (say) and $v_{n+1} = v_n = v_0$. For very small perturbations around (x_0, v_0) the function F in (4.4) may be expanded as a Taylor series, and only the linear terms kept. In chapter 2 we expressed the time derivatives as linear functions of the co-ordinates, 22 but here we are not following the motion in detail; there is no merit in working in terms of local rates of change. Instead we shall express $x_n - x_0$ as X_n, $v_n - v_0$ as Y_n, and assume X and Y to be small enough for the linear terms of the Taylor series to describe their development adequately. Thus we write

$$X_{n+1} = a_1 X_n + a_2 Y_n, \quad Y_{n+1} = a_3 X_n + a_4 Y_n,$$

or[16]

$$\begin{pmatrix} X_{n+1} \\ Y_{n+1} \end{pmatrix} = \begin{pmatrix} a_1 & a_2 \\ a_3 & a_4 \end{pmatrix} \begin{pmatrix} X_n \\ Y_n \end{pmatrix}. \tag{4.5}$$

It is now our task to find what types of behaviour it is possible for X_n and Y_n to exhibit in subsequent samplings. This cannot be achieved without a certain amount of algebraic manipulation, and in the following exercise I lay out the scheme of the analysis, with all the results we shall need, and leave you to fill in the details if you are so minded.

Exercise 4.5. (a) Write

$$\begin{pmatrix} a_1 & a_2 \\ a_3 & a_4 \end{pmatrix} \text{ as A}$$

and

$$\begin{pmatrix} \gamma \cos \theta & (1/\gamma) \sin \theta \\ -\gamma \sin \theta & (1/\gamma) \cos \theta \end{pmatrix} \text{ as C,}$$

whose inverse C^{-1} is

$$\begin{pmatrix} (1/\gamma)\cos\theta & -(1/\gamma)\sin\theta \\ \gamma\sin\theta & \gamma\cos\theta \end{pmatrix}.$$

Form $B = C^{-1}AC$ and show that θ and γ may be chosen so that, depending on the magnitude of the as, B can be written as

$$\begin{pmatrix} b_1 & b_2 \\ b_2 & b_1 \end{pmatrix} \quad \text{or as} \quad \begin{pmatrix} b_1 & b_2 \\ -b_2 & b_1 \end{pmatrix}.$$

b_1 and b_2 may be either positive or negative, but no restriction is involved by assuming, as we shall, that b_2 is positive. Show further that the trace T and determinant Δ are unchanged in transforming A into B, i.e. $a_1 + a_4 = 2b_1$, and $a_1 a_4 - a_2 a_3 = b_1^2 \mp b_2^2$. The physical meaning of this transformation is that we have rotated the axes through θ and afterwards applied a scale distortion, increasing the X co-ordinates by a factor γ and correspondingly decreasing the Y co-ordinates by the same amount. In the new co-ordinate system, with X transformed to $\xi = \gamma(X\cos\theta + Y\sin\theta)$ and Y transformed to $\eta = (1/\gamma)(Y\cos\theta - X\sin\theta)$,

$$\begin{pmatrix} \xi_{n+1} \\ \eta_{n+1} \end{pmatrix} = \begin{pmatrix} b_1 & b_2 \\ \pm b_2 & b_1 \end{pmatrix} \begin{pmatrix} \xi_n \\ \eta_n \end{pmatrix}. \tag{4.6}$$

(b) Take the case of negative sign first, and write $b_1 = s\cos\phi$, $b_2 = s\sin\phi$, where $s^2 = b_1^2 + b_2^2$. Hence show that (4.6) describes the process of rotating the vector (ξ_n, η_n) through ϕ and multiplying its length by s to produce (ξ_{n+1}, η_{n+1}). Continued application of this operation generates a discrete set of points lying on an equiangular spiral, working inwards if $s < 1$, outwards if $s > 1$. This behaviour is analogous to focal stability and focal instability. An example of the latter is shown in fig. 4.11, with $b_1 = 0.8$, $b_2 = 0.7$. The points are joined with straight lines to make the diagram clearer, but the lines have no physical significance.

(c) Show that on a Δ–T diagram, like fig. 2.9, the case of negative sign in (4.6) describes points lying inside the parabola $\Delta = T^2/4$, and that when $\Delta < 1$ the system is stable. The regions of focal stability and instability will be found displayed in fig. 4.12(*d*).

(d) Now turn to the case of positive sign in (4.6), which applies to all points outside the parabola $\Delta = T^2/4$. First give the axes a further twist of 45° by defining the matrix

$$D = \frac{1}{\sqrt{2}} \begin{pmatrix} 1 & 1 \\ -1 & 1 \end{pmatrix},$$

and its inverse

$$\mathbf{D}^{-1} = \frac{1}{\sqrt{2}} \begin{pmatrix} 1 & -1 \\ 1 & 1 \end{pmatrix},$$

and forming $\mathbf{D}^{-1}\mathbf{B}\mathbf{D}$, where $\mathbf{B}$ is the matrix

$$\begin{pmatrix} b_1 & b_2 \\ b_2 & b_1 \end{pmatrix}.$$

This diagonalizes $\mathbf{B}$ so that

$$\begin{pmatrix} \xi'_{n+1} \\ \eta'_{n+1} \end{pmatrix} = \begin{pmatrix} b_1 - b_2 & 0 \\ 0 & b_1 + b_2 \end{pmatrix} \begin{pmatrix} \xi'_n \\ \eta'_n \end{pmatrix}.$$

or

$$\xi'_{n+1} = (b_1 - b_2)\xi'_n \quad \text{and} \quad \eta'_{n+1} = (b_1 + b_2)\eta'_n. \tag{4.7}$$

Hence ξ' and η' evolve independently by geometric progression,

$$\xi'_n = \xi'_0(b_1 - b_2)^n \quad \text{and} \quad \eta'_n = \eta'_0(b_1 + b_2)^n. \tag{4.8}$$

There are many more variants of unstable behaviour than the nodal and saddle-point forms of fig. 2.9, but they involve no new principles. The complications arise because of the two exponents in (4.8) none, one or two may have magnitudes greater than 1. If they are both positive, neither being greater than 1 causes stable convergence to a stationary point, analogous to nodal stability, while both being greater than 1 is analogous to nodal instability. In between we have the analogue of saddle-point

Fig. 4.11. The analogue of focal instability in a discrete sampling. Only the points, not the lines joining them, are significant.

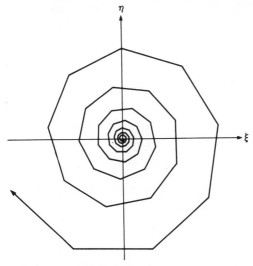

Fig. 4.12. (*a*) Classification of the varieties of behaviour possible with a symmetric *b*-matrix in (4.6), mapped on a *b*-plane. The numbering of the lines and lettering of the domains correspond to the remapping of the same information in (*d*). (*b*) Development according to an unsymmetrized matrix, just outside the stable regime. As in fig. 4.11, only the discrete points, not the lines joining them, are significant. (*c*) the same as (*b*), but transformed into a symmetrical *b*-matrix corresponding to the point *P* in (*a*). (*d*) Remapping of (*a*) onto a Δ–T plane. This also includes the regions of focal stability and instability described by an antisymmetrical *b*-matrix.

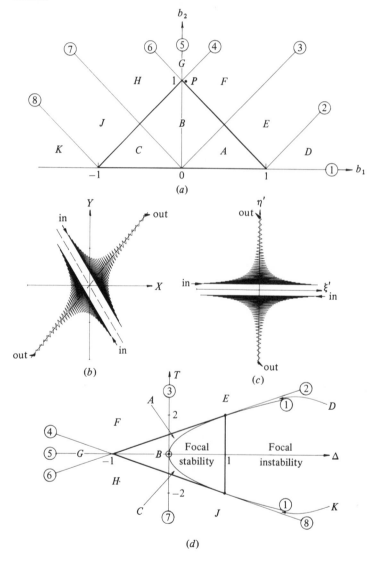

instability. In addition to this, we must consider the possibility that the exponents may be negative, leading to alternation of sign being super-imposed on the behaviour. Let us first enumerate all the cases in terms of the size of b_1 and b_2. This is most readily presented as a diagram (remember that we are considering only the case of positive sign in (4.6), and taking b_2 as positive. It is obvious from (4.8) that if b_2 were negative the roles of ξ' and η' would be interchanged, as by rotation of the axes through $90°$, but the stability considerations would be exactly the same.

In fig. 4.12(a) the b_1–b_2 plane is divided into ten regions, in each of which the system behaves differently. For example, in F, $(b_1 - b_2)$ lies between -1 and 0, while $(b_1 + b_2)$ is greater than 1; hence ξ' alternates in sign as it decays to zero, while η' keeps the same sign and grows by geometric progression. This is illustrated in fig. 4.12(b) and (c) for $b_1 = 0.05$, $b_2 = 0.9912$, as well as for an untransformed A-matrix

$$A = \begin{pmatrix} 0.2 & 0.4 \\ 2.4 & -0.1 \end{pmatrix}$$

which has very nearly the same $T(0.1)$ and $\Delta(-0.98)$, lying close to the stability limit, as shown by the point P in the diagram. One can see the independent solutions superimposed; if we were to start the system at a point on either broken line in (b), or on either axis in (c), it would continue on that line, either convergently alternating in sign, or diverging in a straight path from the origin. The envelope of the points shows the analogy to saddle-point instability. From fig. 4.12(a), and with the help of (4.7) or (4.8), we can read off the behaviour in each of the ten regions, with the result presented in Table 4.1. F and H occupy the same cell; for the region F the convergent solution alternates while the divergent does not: for H it is the divergent solution that alternates while the convergent does not.

Finally, to make the whole scheme apply to the general A-matrix of (4.5), we translate each line in fig. 4.12(a) into a relation between Δ and T, and replot the lines on a Δ–T diagram, fig. 4.12(d). The topological

Table 4.1

	both alternating	one alternating	neither alternating
both convergent	C	B	A
one convergent ⎫ one divergent ⎬	J	$F\ H$	E
both divergent	K	G	D

structure is unaltered, and each labelled region may be identified by the lines which bound it. Everything inside the central triangle represents stability, and everything outside some form of instability. The origin represents the same state in both diagrams, and is one of superstability, where $b_1 = b_2 = 0$ and hence, from (4.7), the system collapses immediately into the stationary state. This means that to discuss convergence to the stationary state we need to go beyond the linear terms of the Taylor expansion, and will find that convergence does not follow a geometrical progression. We shall meet an example presently.

Exercise 4.6. Derive expressions for the rate of decrement of the two modes in the region C, starting with (4.8), and remembering that $b_1 - b_2$ and $b_1 + b_2$ are both negative, $-\lambda_1$ and $-\lambda_2$, say. Show that λ_1 and λ_2 are the solutions of $\lambda^2 + T\lambda + \Delta = 0$. Hence show that if the representative point is only just within the triangle of stability, so that $T = -(1 + \Delta) + q$, and $q \ll 1$, then $\lambda_1 \sim 1 - q/(1 - \Delta)$ and $\lambda_2 \sim \Delta[1 + q/(1 - \Delta)]$.

The value of n in (4.8) needed to reduce the amplitude of a mode by a factor e, the equivalent of the time-constant, is $-1/\ln \lambda$, i.e. $\sim (1 - \Delta)/q$ or $-1/\ln \Delta$. The former rises to infinity like $1/q$ as the critical line is approached, and one mode goes from stability to instability. The other mode remains quite heavily damped unless Δ is very close to unity.

Application of stability test to the impact oscillator

The simplest stationary state of the model shown in fig. 4.9 is that whose phase–space representation is fig. 4.10(*a*). Let us choose to sample the position and velocity at such times that $\omega t = n\pi$, n being an integer. Note that we sample twice per cycle, so that if the state of the ball is (x_0, v_0) at $t = 0$, the stationary state we seek finds it at $(-x_0, -v_0)$ when $\omega t = \pi$. The reverse path mirrors the forward path, but this is only possible if the walls are identical, with the same coefficient of restitution, ε. This is a point we shall return to. If the frequency of the wall oscillation lies in the audible range, the note produced by the collisions will be at twice this frequency, an octave above.

Exercise 4.7. (a) When $t = n\pi/\omega$, let the ball (of negligible size) be x_n to the right of the centre line of the diagram, which is also midway between the walls at this instant, and let it move to the right with velocity v_n. Show that it hits the right-hand wall at t_n such that

$$Y_n Z_n = 1 - X_n + A \sin Z_n, \tag{4.9}$$

where $Z_n = \omega t_n$, $X_n = x_n/l$, $A = a/l$ and $Y_n = v_n/\omega l$. Hence show that, following Newton's law of collision, it rebounds with velocity v_{n+1} to the left, such that

$$Y_{n+1} = \varepsilon Y_n - (1+\varepsilon)A \cos Z_n. \tag{4.10}$$

Finally, show that at the next instant of sampling, $t = (n+1)\pi/\omega$, it has reached a point x_{n+1} to the left of the centre line, such that

$$X_{n+1} = (\pi - Z_n)Y_{n+1} - 1 - A \sin Z_n. \tag{4.11}$$

(4.9)–(4.11) are the basic equations of the problem. The only one that presents any difficulty is (4.9) which requires solution by successive approximation, and is time-consuming on a small calculator.

(b) We therefore replace strictly sinusoidal shaking of the walls by a succession of parabolic motions, as illustrated in fig. 4.13. This changes the numerical results in detail but leaves the essential qualitative behaviour unaltered. The appropriate modification that yields the same wavelength and amplitude is

for $\quad 0 < Z < \pi, \quad \sin Z \rightarrow (4/\pi)(Z - Z^2/\pi),$

for $\quad -\pi < Z < 0, \quad \sin Z \rightarrow (4/\pi)(Z + Z^2/\pi),$

and similarly for all successive cycles. Show that (4.9)–(4.11) now take the form

$$Y_n Z_n = 1 - X_n + (4A/\pi)(Z_n - Z_n^2/\pi), \tag{4.12}$$

$$Y_{n+1} = \varepsilon Y_n - (4A/\pi)(1+\varepsilon)(1 - 2Z_n/\pi), \tag{4.13}$$

$$X_{n+1} = (\pi - Z_n)Y_{n+1} - 1 - (4A/\pi)(Z_n - Z_n^2/\pi). \tag{4.14}$$

The transcendental equation (4.9) is now replaced by a simple quadratic equation (4.12), with something like tenfold increase of computing speed. By substituting (4.12) in (4.13) and (4.14), and rearranging, (X_{n+1}, Y_{n+1}) can be expressed as some function

Fig. 4.13. $f(Z) = \sin Z$ (curve a) and $(4/\pi)(Z \mp Z^2/\pi)$ (curve b).

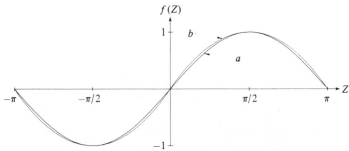

$F(X_n, Y_n)$, thus realizing the mapping function (4.4) for this problem. It is usually easier to proceed without doing so, solving (4.12)–(4.14) in sequence.

(c) In the stationary state, $X_{n+1} = X_n = X_0$, $Y_{n+1} = Y_n = Y_0$, $Z_n = Z_0$. Show that

$$Z_0 = \pi\{[(1 + \varepsilon^2)/2(1 - \varepsilon)^2 + 1/4A]^{\frac{1}{2}} - \varepsilon/(1 - \varepsilon)\}$$
$$Y_0 = -(4A/\pi)(1 + \varepsilon)(1 - 2Z_0/\pi)/(1 - \varepsilon) \qquad . \qquad (4.15)$$
$$X_0 = Y_0(\pi/2 - Z_0)$$

A typical example of a stationary state is shown in fig. 4.14, with the nth and $(n+1)$th sampling points shown in equivalent positions (X_0 is negative). The collisions with the walls at A and B are such that the wall motion compensates for the less-than-elastic collision. Thus A lies between $\pi/2$ and $3\pi/2$. In fact it must be between $\pi/2$ and π for a stable state, as may be seen by considering a collision at C. Even if C and its equivalents satisfy (4.15), the resulting stationary state must be unstable; for if the ball happens to arrive at C a little late, it will rebound rather more slowly and arrive for its next collision even later. By contrast, late arrival at A finds the ball struck more sharply, and it is possible that it will be more punctual for its next encounter. This needs investigation, and is of course where the stability analysis comes in.

(d) Consider small deviations of X, Y and Z from their stationary values, writing $\xi_n = X_n - X_0$, $\eta_n = Y_n - Y_0$, $\zeta_n = Z_n - Z_0$, and taking

Fig. 4.14. A trajectory $X(Z)$ (equivalent to $x(t)$) for the impact oscillator locked in a fundamental mode.

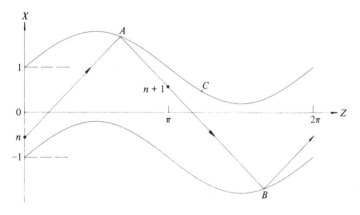

note only of linear terms in ξ and η. Thus by differentiating (4.12), one finds

$$Y_0\zeta_n + Z_0\eta_n = -\xi_n + (4A/\pi)(\zeta_n - 2Z_0\zeta_n/\pi).$$

Treat (4.13) and (4.14) similarly, and sort out the resulting expressions with the help of (4.15) to give

$$\left.\begin{array}{l} \xi_{n+1} = [\varepsilon - \alpha(\pi - Z_0)]\xi_n + [\pi\varepsilon - \alpha Z_0(\pi - Z_0)]\eta_n \\ \eta_{n+1} = -\alpha\xi_n + (\varepsilon - \alpha Z_0)\eta_n \end{array}\right\} \quad (4.16)$$

where

$$\alpha = 4A(1+\varepsilon)^2/\pi^2 Y_0.$$

The coefficients in (4.16) form the A-matrix of (4.5), and T and Δ are quickly evaluated:

$$T = 2\varepsilon - \pi\alpha \quad \text{and} \quad \Delta = \varepsilon^2. \quad (4.17)$$

This simple result shows that as α, defining the amplitude of wall motion, is altered the representative point in fig. 4.12(*d*) moves vertically along the line $\Delta = \varepsilon^2$, and we may expect a range of stability. The highest amplitude A_{c0} at which the ball may be captured in this particular orbit is determined by the crossing of the line $\Delta + T + 1 = 0$, line 4 in fig. 4.12(*d*). Here $Y_0 = 4A/\pi$ and, from the second equation of (4.15), $Z_0 = \pi/(1+\varepsilon)$. From the first equation of (4.15) this occurs when

$$A_{c0} = (1+\varepsilon)^2/2(1+\varepsilon^2). \quad (4.18)$$

From Ex. 4.6 we expect the approach to the stable orbit, just inside the triangle of stability, to be described by two independent modes of alternating sign; one should decay rapidly as Δ^n, i.e. ε^{2n}, and the other slowly, at a rate proportional to distance from the critical condition. This is confirmed by computation. If you wish to test this for yourself, e.g. by computing successive η_n from (4.12)–(4.15), you will find it helpful to know that the two independent modes lie along the lines $\eta = 2.8525\xi$ (rapidly decaying) and $\xi \sim 0$ (slowly decaying).

As for the lowest amplitude at which the stable orbit can be found, it is easy to fall into the trap of assuming that it is determined by the crossing of line 6, $\Delta - T + 1 = 0$, only to discover that this demands a negative value of α, contrary to hypothesis. In fact, the first equation of (4.15) shows that when $A = (1-\varepsilon)/2(1+\varepsilon)$, $Z_0 = \pi$. At this value of Z_0 the walls have the greatest velocity and can only just sustain the speed of the ball, $Y_0 = 2/\pi$, that will keep it bouncing in synchronism. The transition as A falls below this value resembles the hysteretic jumps in fig. 4.4, in that the solution adopted hitherto ceases to exist beyond the critical point. By contrast, at the upper limit A_{c0} the solution still exists, but ceases to be stable. To anticipate once more the terminology of chapter 5, on the low side there is a limit point in stability, on the high side (as we shall see) a *stable*   132 *symmetric transition.*

Behaviour outside the range of stability; subharmonics

When $A < (1-\varepsilon)/2(1+\varepsilon)$ we must consider the possibility that the system is excited in a subharmonic mode, one in which the ball moves so slowly that it requires more than one cycle of oscillation to complete its to-and-fro passage. The way in which this may be achieved is shown in fig. 4.15, where trajectory 1 is like fig. 4.14, except that the vibration amplitude has been considerably reduced, while trajectory 3 represents a complete cycle which takes three times as long. If each collision is audible, the fundamental note of trajectory 1 sounds an octave above the wall frequency ω; the third subharmonic of trajectory 3 is a fifth below ω, and the fifth subharmonic an octave and a major third below ω. Collisions with the walls occur in similar circumstances as for the earlier calculation, and it is easy to see that (4.12)–(4.14) need to be changed only by replacing $(\pi - Z_n)$ in (4.14) by $(m\pi - Z_n)$, where m is an odd integer. For $m = 3$, Z is measured from O', for $m = 5$ from O'', and so on. The calculations of Ex. 4.6 proceed with similarly minor alterations to give the equivalent of (4.16):

$$\xi_{n+1} = [\varepsilon - \alpha(m\pi - Z_0)]\xi_n + [m\pi\varepsilon - \alpha Z_0(m\pi - Z_0)]\eta_n$$

and

$$\eta_{n+1} = -\alpha\xi_n + (\varepsilon - \alpha Z_0)\eta_n \qquad \qquad , \qquad (4.19)$$

from which it follows that

$$T = 2\varepsilon - m\pi\alpha \quad \text{and} \quad \Delta = \varepsilon^2,$$

Fig. 4.15. The same as fig. 4.14, extended to include third and fifth subharmonics.

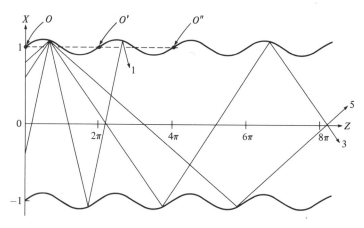

differing only from (4.17) by the presence of $m\pi\alpha$ instead of $\pi\alpha$ in T. The upper limit for A that will permit the mth subharmonic is now given by the condition $\Delta + T + 1 = 0$ or $Y_0 = 4mA/\pi$. This, in its turn, gives $Z_0 = \pi[1 + m(1 - \varepsilon)/(1 + \varepsilon)]$ at the upper critical point, and

$$1/A = m^2[2 + (1 - \varepsilon)^2/(1 + \varepsilon)^2] - 1. \tag{4.20}$$

The lower critical point, as before, marks the point where $Z_0 = \pi$, which occurs when $A = (1 - \varepsilon)/2m(1 + \varepsilon)$. In contrast to the fundamental response which can be stable for any value of ε, though only over a narrow range of A when ε is small, the subharmonics are excited only when ε is rather near unity. Fig. 4.16 makes this clear.

What, you may ask, happens in the extensive region lying below the limit of the fundamental response, and outside the limits of subharmonic response? Probably chaotic motion, by which we mean motion without periodicity. It is not random, since each sample is uniquely related by something like (4.4) to its preceding sample, but there is nothing to guarantee that the system must settle down to a regular pattern. If one lowers the amplitude of the driver, moving down, e.g., the broken line in fig. 4.16, one may expect a range of fundamental response followed by noise, from which with luck the third subharmonic will emerge. Then a further noisy range leading to the fifth subharmonic, the last to be accessible with this value of ε. We leave it like this, since the search for possible periodic motions other than simple subharmonics is likely to be tedious and of rather slight interest.

106
119

It might indeed be said that this type of strict subharmonic response is

Fig. 4.16. The domains of ε and A in which various subharmonics are stable. Note the logarithmic plot of A.

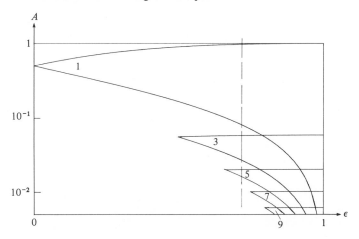

not worth spending much time on, since examples in physics and engineering, apart from contrived examples like this one, are distinctly uncommon. A more common type of subharmonic response is one in which the system responds very nearly in synchronism with the driving force, but with relatively minor variations from one cycle to the next, so that the fundamental Fourier component of the response is indeed a subharmonic of the driver, but there is present a strong Fourier component at the driving frequency. This is what we shall find happens above the critical values of A given by (4.20), and as this exemplifies a surprising and interesting type of physical behaviour we shall devote considerable time to it. But we must start further back, and before dealing with mapping problems of the type (4.4) we will look carefully at one-parameter mapping which presents most of the essentials in a readily visualized form.

Bifurcation and chaos in one variable

Imagine that we generate a sequence of numbers $x_1, x_2, \ldots, x_n,$ $\ldots$ by the process of operating on each successive number in accordance with a strict rule to obtain the next in the sequence:

$$x_{n+1} = f(x_n). \tag{4.21}$$

The long-term behaviour of the sequence will depend on the function f. Thus if $x_{n+1} = ax_n$, the sequence will converge to zero for $a < 1$ and diverge for $a > 1$. Divergence can be prevented by limiting the range of f, as in fig. 4.17, where the peak value of f is less than unity.[17] It is readily seen that successive values are obtained by walking up the staircase bounded by f and by the line $x_{n+1} = x_n$. As drawn, the sequence converges on the intersection, but in oscillatory fashion. Obviously the sequence can settle down to a steady value only at such an intersection, and the first point to examine is whether the intersection gives a stable end-point.

Exercise 4.8. (a) Let the curve (4.21) intersect the line $x_{n+1} = x_n$ at the value x_0, and expand it as a Taylor series around this point:

$$x_{n+1} - x_0 = \alpha(x_n - x_0), \quad \text{where } \alpha = (df/dx)_{x_0}.$$

Show that successive values of $x - x_0$ form a geometrical sequence and that the intersection is

      unstable   if $|\alpha| > 1$

      neutral   if $\alpha = \pm 1$

      stable   if $|\alpha| < 1.$

In the range $0 < \alpha < 1$, successive points converge on x_0 while remaining on the same side of it, while in the range $-1 < \alpha < 0$ they converge in oscillatory fashion. When $\alpha < -1$, they oscillate from side to side while diverging from x_0.

(b) When $\alpha = 0$, and the intersection happens to lie at the peak of f, the system is said to be superstable. The leading term in the Taylor expansion is quadratic:

$$x_{n+1} - x_0 = -\beta(x_n - x_0)^2, \quad \text{where } \beta = -2(\mathrm{d}^2 f/\mathrm{d}x^2)_{x_0},$$

or

$$\delta_{n+1} = -\beta\delta_n^2,$$

in which δ is written for $x - x_0$. Verify that the solution of this equation takes the form

$$\delta_n = -a^{2^n}/\beta,$$

where a is a constant, less than unity, whose sign is determined by the starting value, $\delta_0 = -a/\beta$. All subsequent δ_n are negative, and the rapider-than-exponential convergence of δ_n to zero is worth observing by computing a few terms in the sequence.

The behaviour becomes interesting when $\alpha < -1$, that is, when the curve for f slopes downwards at the intersection more steeply than 45°.

Fig. 4.17. The curve $x_{n+1} = f(x_n)$ and the line $x_{n+1} = x_n$ to illustrate staircase procedure for generating the sequence x_n, given the initial value x_1.

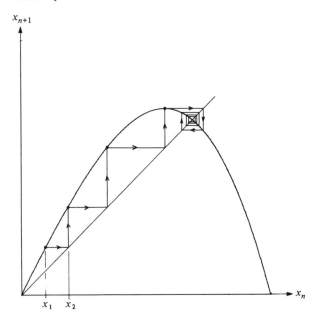

We know that x_n is restricted to the range $0 < x_n < 1$, and that there is no stable value. What then happens to x_n? At this point it is convenient to choose a specific form for f, though it does not matter, within a wide range, what the form is. Much published work is based on the simplest choice $f(x) = Ax(1-x)$, a symmetrical parabola with its peak at $x = \frac{1}{2}$. There is a danger of propagating the belief that the symmetry of the function plays a part in determining some of the essential features of the behaviour. This is not so, but to avoid suspicion we shall choose

$$f(x) = Ax(1-x^2),\tag{4.22}$$

which has a peak value of $2A/3\sqrt{3}$, or $A/2.598\,08$, when $x = 1/\sqrt{3}$. The values of x_n remain confined to the range $0 < x_n < 1$ so long as $A < 2.598\,08$. By changing A the gradient of the curve at the intersection with the $45°$ line can be altered at will. At the intersection, $x = Ax(1-x^2)$, i.e. $x = (1-1/A)^{\frac{1}{2}}$, and the slope is $3 - 2A$; slopes from 1 to -2.196 are possible, and when $A > 2$ the intersection is unstable. Curves for various values of A are shown in fig. 4.18(a).

As A approaches 2 from below, the oscillatory convergence of the sequence becomes very slow, until at this value it virtually stops, and the values of x_n alternate between $x_0 \pm \delta$, with δ changing very slowly, and then only because of the curvature of f. When A is slightly greater than 2 the alternation grows in magnitude, but eventually stabilizes. Just as a feedback oscillator may reach a limit cycle as a consequence of non-linearity in the amplifier, so here a limit cycle is reached, with a period of 2. The two points in 4.18(a), Q and R, lying on f at mirror reflection points with respect to the $45°$ line, represent this stable, period 2, behaviour. The implication is that two applications of the function f define a new function $g(x) = f[f(x)]$ which generates a converging sequence of values: since

$$x_{n+2} = Ax_{n+1}(1-x_{n+1}^2) \quad \text{and} \quad x_{n+1} = Ax_n(1-x_2^n),$$
$$x_{n+2} = A^2 x_n(1-x_n^2)[1-A^2 x_n^2(1-x_n^2)^2].\tag{4.23}$$

The function on the right-hand side is $g(x_n)$, a ninth-degree polynomial. Further analysis demands computer plotting of the curves, and fig. 4.18(b) shows, as curve 2, the results of two iterations; that is, (4.23) with $A = 2.2$. Of the three intersections with the $45°$ line, P is unstable but Q and R, with negative gradients less than $45°$, are stable. These are the two values between which x_n alternates. The three points are labelled in the same way in fig. 4.18(a). The process described here, in which a stable sequence becomes unstable and is replaced by two new stable sequences, is called a *bifurcation*. It reminds one of the way an Euler strut changes from one   14 symmetrical stable configuration to two mirror images on opposite sides

Fig. 4.18. (a) $x_{n+1} = Ax_n(1 - x_n^2)$ for the values of A shown against the curves. For $A = 2.1$, P is an unstable point and the sequence converges to an alternation of Q and R, mirror images with respect to the line $x_{n+1} = x_n$. (b) Curves for one, two and four iterations of the function $x_{n+1} = Ax_n(1 - x_n^2)$.

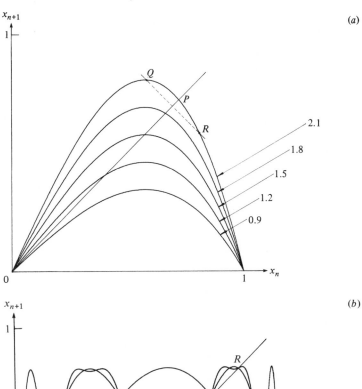

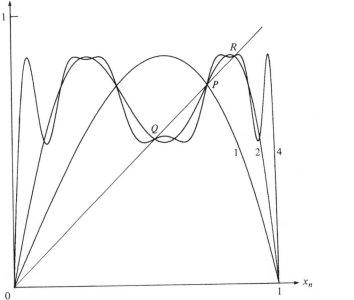

of the original configuration, which becomes unstable. Here also the original sequence remains a possibility, albeit unstable. The two stable sequences can be visualized as mirror images in the 45° line; they are identical but displaced by one term relative to one another.

Inspection of curve 2 in fig. 4.18(*b*) leads one to believe that only a small increase in *A* above 2.2 will make the intersections *Q* and *R* in their turn unstable, so that a second bifurcation will occur, leading to a mapping function $g(g(n))$, or four iterations of *f*, whose algebraic form is an unmanageable polynomial of degree 81. This is shown as curve 4 in fig. 4.18(*b*), for the same value, 2.2, of *A*, from which it is confirmed that *A* is insufficient, though only just, to lead to four stable intersections with the 45° line. Further increase of *A* will cause the slope of curve 4 at *Q* and *R* to become steeper until it exceeds 45° and each single intersection becomes three, of which two are stable. The sequence x_n now has period 4, and as *A* is increased bifurcation succeeds bifurcation, with period doubling each time, so that the steady state of the sequence contains a repeat pattern of numbers containing successively 2, 4, 8, 16, ..., 2^n, ... terms. Here are a few examples, with the repeat pattern underlined:

$A = 1.6$ $\underline{0.612\,372\,4}, 0.612\,372\,4, \ldots$

    2.15 $\underline{0.826\,731\,7, 0.562\,596\,4}, 0.826\,731\,7, \ldots$

    2.27 $\underline{0.873\,565\,9, 0.469\,736\,9, 0.831\,020\,1,}$

        $\underline{0.583\,667\,5}, 0.873\,565\,9, \ldots$

It is rather a tedious business finding the critical values of *A* at which successive bifurcations occur because, as we noted with the first, when *A* is close to the critical value the convergence or divergence is very slow. When one looks for the first or second bifurcation it is no great matter to let the computer churn out successive iterations until the pattern stabilizes, and to adjust *A* as closely as one likes. But finding the eighth bifurcation, from a periodic sequence 128 steps in length to one of 256 steps, takes a well-nigh intolerable time. It is a good deal easier to look for superstable points instead. The argument can be followed with the help of 94 fig. 4.18. Consider *f* itself first; as *A* is increased (see fig. 4.18(*a*)) the 101 superstable case $A = 1.5$ obviously lies between the regime of positive and of negative slopes. Similarly, if we imagine how curve 2 in fig. 4.18(*b*) evolves with *A*, there will at first ($A < 2$) be only one intersection at *P*; at the point of bifurcation curve 2 will be tangential and points *Q* and *R* will spread out from *P*. Only at a slightly later point will the intersections at *Q* and R be at extrema of curve 2; this is the moment when the system is superstable. Then the slope at the intersection becomes negative and eventually the next bifurcation is reached. This is a general pattern – between successive bifurcations there is a superstable point. And we now note that superstable points are easily located, for one of them always coincides with the peak of f ($x = 1/\sqrt{3}$ in this case); if $df/dx = 0$, so also is $(d/dx)[f(f(x))]$. To find the value of A at which the sequence is superstable for n iterations, it is only necessary to put $x_1 = 1/\sqrt{3}$ and compute the result of applying $f(x)$ n times; A is then adjusted until the result is $1/\sqrt{3}$, and this is the required value. Thus for the choice (4.22) of f, the following set of values is obtained; it will be recalled that between each successive pair there is a critical value at which a bifurcation occurs, so that Table 4.2 shows immediately how the bifurcations get closer together:

Table 4.2

number of iterations	A for superstability	$\Delta A \times 10^{10}$	ratio of successive ΔA
2^0	1.5		
		623 120 344	
2^1	2.121 320 344		4.385 709
		141 669 311	
2^2	2.262 989 655		4.591 654
		30 853 659	
2^3	2.293 843 314		4.652 670
		6 631 388	
2^4	2.300 474 702		4.665 628
		1 421 328	
2^5	2.301 896 030		4.668 43
		304 455	
2^6	2.302 200 485		4.669 1
		65 207	
2^7	2.302 265 692		4.669 (7)
		13 964	
2^8	2.302 279 656		

It was first observed by Feigenbaum,[18] and extensively studied by him and others afterwards, that the separation of critical values of A became smaller by a factor that tended to a constant. This in itself would have aroused mild interest, but the activity that followed Feigenbaum's announcement sprang from his recognition that the same constant always appears when f has a quadratic maximum and gives rise, as is normal, to a bifurcation sequence. Whatever the shape, subject to these conditions, the gaps between bifurcations decrease by a factor that tends to a universal constant, 4.669 201 6 . . . . Obviously the computation must be carried to many more decimal places to verify this number in any given case. It is also clear that this result would never have been discovered without the use of a computer or, at least (as Feigenbaum has insisted), a programmable pocket calculator. We have here a perfect example of the computer used as an experimental tool.

No one should conclude from the universality of Feigenbaum's result that it is fairly obvious, or that it can be proved in an elementary manner and the value of the constant derived straightforwardly. For all the effort he has put into making the argument accessible, the fact remains that no elementary derivation has yet been given, and Feigenbaum's is quite formidable; only real mathematicians should attempt to master it.

What is even more remarkable about Feigenbaum's universality is that it is not confined to one-parameter mapping. We find it also in two-parameter mapping, such as the impulse oscillator to which we shall soon return, and even in what we may describe as real-life problems, of which a few have been studied closely enough to satisfy the experimenters that the observed bifurcations follow the Feigenbaum scheme. There is, in fact, every justification for treating it very seriously as a generalization of many different situations. And the fact that only rather sketchy arguments can be adduced for the appearance of the same constant in every one of them only serves to make the study more interesting as an example, all too rare in classical physics nowadays, of a phenomenon which was entirely unsuspected before its discovery.

On the assumption that successive bifurcations, and therefore period doublings, will continue to be 4.669 . . . times closer together, we infer from Table 4.2 that when A reaches 2.302 283 462 the periodicity of the sequence will have become infinite, although this value of A is still well below the value 2.598 08 at which containment breaks down. What happens is that the sequence of numbers becomes perfectly aperiodic, so that in the course of time whole stretches of the available number-range are visited and densely occupied. This behaviour is called *chaotic*, and one must be clear that there is a firm distinction to be drawn between this and random behaviour. In a random sequence of numbers each successive term is governed by a probability distribution, but is not exactly determined; thus a given row of terms may occur on many occasions, but each time followed by a different number. In the chaotic sequence as just introduced each term is uniquely determined by the preceding one. The idea of chaos may be generalized, for example by making each term depend on the two preceding it, but it is of the essence of chaos, as used technically, that it is completely aperiodic and completely determinate.

In spite of this determinacy, chaos and randomness have this in common, that the chaotic sequence of numbers has no long-term predictability. This is best illustrated by studying the effect of a small perturbation on the subsequent behaviour, as in fig. 4.19. The starting values differ by hardly more than 10^{-7}, and for about 40 steps the sequences follow the same course nearly enough to be hardly distinguishable by eye. The computer numbers, however, are diverging in a very roughly exponential fashion, the difference doubling every 2–3 steps, and by the end of the graph this divergence has multiplied until it dominates the behaviour. From then on the sequences follow completely independent paths. One sees from this example that chaos does not destroy predictability entirely – if the initial value is known fairly well, the behaviour can be predicted for a few terms, and the better the specification of the initial value, the more terms for which prediction is possible. But the 'exponential' divergence carries the implication that doubling the accuracy of the initial specification only lengthens the range of prediction by two or three steps. If we are thinking of a real physical situation in which chaos occurs, the initial state is likely to be determined by a measurement, and we must reconcile ourselves to accepting that increasing the accuracy of this measurement pays very small dividends. Ultimately, and sooner rather than later, the exponential growth of an initial error will destroy any hope of predicting the outcome.

In a real physical system pure chaos is never found. There is always some background noise, whether thermal or some larger disturbance from man-made or other vibrations. This has the same effect as a measurement error, adding a random increment to each successive x_n, which will be

Fig. 4.19. The sequence x_n for $A = 2.4$: $x_1 = 0.8$ for sequence a and 0.800 000 1 for b.

amplified as in fig. 4.19. The range of prediction cannot be longer than the number of steps possible before a typical noise increment is amplified to the point of dominating subsequent behaviour; unlike a measurement error which in principle can be reduced, there is a lower limit to the noise level, even under ideal conditions, and consequently an upper limit to the range of prediction.*

The example illustrated in fig. 4.19 is well in the chaotic regime, with $A = 2.4$. Just above the value of A (2.302 28 ...) at which chaos sets in, when $A = 2.305$, say, the divergence is much slower, taking about 20 steps to double a small initial difference. This is what one would expect, since when A is less than the critical value for chaos, the sequence settles down to a regular pattern with a period of 2^n. Differences in the initial value, provided they are small, will not change the ultimate pattern and the system is stable in this sense, in contrast to the instability of the chaotic pattern. Too big a change of the initial value may cause the pattern to slip a few notches, settling down to the same sequence but displaced by a number of steps, i.e. to one of the 2^n 'mirror images' of the basic sequence.

> *Exercise 4.9.* Use a programmable pocket calculator to investigate the period-doubling of the function $x_{n+1} = Ax_n(1 - x_n)$; in particular, find the superstable values of A and check Feigenbaum's result.

Further properties of the one-variable map

It will soon become clear that even the innocuous-looking one-variable system possesses an extraordinary potentiality for complicated behaviour. I shall make no attempt to describe anything more in detail, but illustrate some of the phenomena by reference to our original example, $x_{n+1} = Ax_n(1 - x_n^2)$. First, let us look at what we have already accomplished from a different viewpoint, concentrating still on the superstable points. If we set $x = 1/\sqrt{3}$ and find after performing m iterations that x_{m+1} is $1/\sqrt{3}$ again, we have discovered a value of A that generates a superstable cycle of m terms. Fig. 4.20(a) shows how x_{m+1} varies with A for various choices of m ($= 1, 2, 4, 8$); the intersections with the horizontal line $x_m = 1/\sqrt{3}$ determine the superstable points. $A = 1.5$ is superstable for $m = 1$, and automatically for all higher values of m, not solely $m = 2^n$. We are not surprised to find all the curves crossing the line $x_{m+1} = 1/\sqrt{3}$

---

* In most physical measurements it is thermal noise that sets the limit to precision. At very low temperatures, or in special circumstances, thermal noise may become irrelevant, and the limit be set by the uncertainty principle, which is inescapable.

where $A = 1.5$, though it may require a little thought to appreciate why they are all tangential at the crossing. The curve for $m = 1$ never crosses this line again, but that for $m = 2$ crosses once more where $A = 2.121 \ldots$, the superstable point for 2 iterations, and therefore for all multiples of 2. Similarly where $A = 2.26 \ldots$ the curves for $m = 4$ and all multiples of 4 cross the line $x_m = 1/\sqrt{3}$. These successive crossings mark the points we have already determined by direct computation in illustrating Feigenbaum's results.

Fig. 4.20. In all these curves, $x_1 = 1/\sqrt{3}$, and the curves show $x_{m+1}(A)$ for various values of m, the number of iterations.

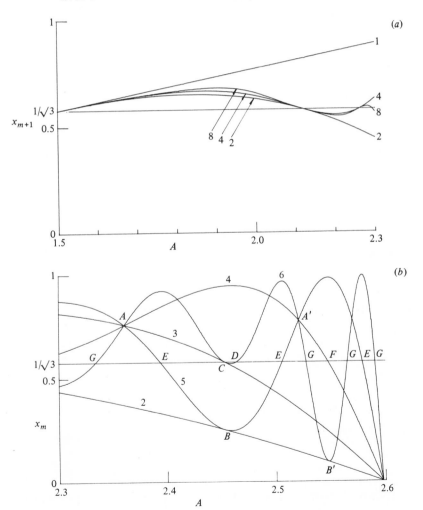

When the curve-plotting is carried into higher values of A, as in fig. 4.20(b), and other values of m filled in, a wealth of superstable points is revealed. You may care to amuse yourself working out why the lines are concurrent at certain points like A and A', or tangential (though not on the line $1/\sqrt{3}$) as at B and B', or tangential on the line, as at C (if you understood the last paragraph you will have no problem with this one). What is more immediately relevant is that at C ($A = 2.453$) there is a superstable point for a 3-fold cycle. As A approaches 2.453, the chaotic state is suspended, to be replaced by a short stretch in which the regular 3-cycle is stable. How narrow this interruption is can be gauged from fig. 4.21, showing a small portion of the 3-fold iteration around $x = 1/\sqrt{3}$. The middle curve, which crosses the $45°$ line with horizontal tangent, is the superstable curve corresponding to C in fig. 4.20(b). Before D is reached, when $A = 2.4608$ in fact, the 3-fold iteration crosses the $45°$ line with negative gradient of -1. This is the value of A at which the 3-fold cycle bifurcates, and just beyond it is D, at which the 6-fold cycle is superstable. As A continues to rise, the pattern of Feigenbaum bifurcation develops, with successive cycles of 3×2^n terms, until very soon chaos is come again.[19]

At the lower edge of its band (curve a in fig. 4.21) the 3-cycle vanishes in a quite different way, analogous to a limit point instability. Curve (a) in fig. 4.22 shows the cycle at its limit, $A = 2.4505$, the three alternating values being joined to show the pattern clearly. In the following three diagrams only every third point is plotted, and (b) shows 200 3-fold iterations with $A = 2.4504$. There has been an abrupt transition to an aperiodic state, but one which retains a strong memory of the lost regularity. It lingers near the pattern of the 3-cycle, but of course cannot be captured, and after a stretch of exploration finds the 3-cycle once more, though not necessarily after $3n$ iterations, so that in this representation it is not always the same point that appears. This phenomenon of *intermittency*[20] is very irregular in the present example, though not always. The irregularity here has its origin in the wide jumps from one iteration to the next, so that the point reached in an iteration is very sensitive to the initial value. Thus the two long sequences marked as S_1 and S_2 in (b) show in detail a slow drift of x, but not exactly the same values. Here are a few representative values, taken from the middle of the sequences:

S_1	0.591 964,	0.592 072,	0.592 180,	0.592 289,	0.592 401
S_2	0.591 979,	0.592 087,	0.592 195,	0.592 305,	0.592 417.

The displacement of the two sequences by as little as 0.000 015 is enough to make the difference between the patterns after the loitering is ended. Fig. 4.22(c) and (d) show that as A is further reduced the intermittency gets more irregular, and more typically chaotic.

By contrast, two maintained oscillators coupled just too weakly for entrainment to occur show a very regular intermittency. The circuit of fig. 4.22(e), from which the feedback maintenance is omitted (see fig. 2.3), 13 employs a resistor to couple the oscillations. This is the simplest case to analyse, but I shall only quote the result here:[21] if the two circuits oscillate at the same frequency but with a variable phase difference (another way of saying they do not oscillate at the same frequency!), the phase difference ϕ develops according to the equation

$$\dot{\phi} = \varepsilon + \sin \phi,$$

in which ε is proportional to the coupling resistor. The unit of time has been chosen to express the equation in the simplest terms. When $\varepsilon < 1$, a value of ϕ can be found such that $\dot{\phi} = 0$, and the circuits are entrained. But when ε is slightly greater than 1 entrainment fails; however, as fig. 4.22(f) shows, ϕ changes only very slowly around the lost point of entrainment, $\sin \phi = -1$.

It cannot be expected that intermittency should be a common phenomenon. In the first place, when the representative point (e.g. x_n in the

Fig. 4.21. Portion of the curves for three iterations, showing the limits of stability a and c, and the superstable mapping b: $A = 2.4505$ for a, 2.4530 for b and 2.4608 for c.

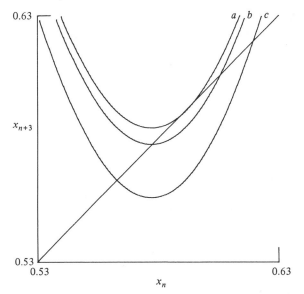

mapping problem, ϕ for the coupled oscillators) fails to find a point of stability, there must be no other available to capture it, such as the 2-cycle when A rises above 2. Secondly, it must not have too far to wander before returning to its point of near-capture. In these two examples the region available for wandering is one-dimensional and of finite extent – ideal for intermittency. When we turn to the non-linear driven oscillator, the mode

Fig. 4.22. Intermittency: (a) $A = 2.4505$, as in curve a of fig. 4.21. The regular 3-fold sequence is on the limit of stability. (b) Every third point is shown, for $A = 2.4504$. The sequence loiters near the stable points in (a) but cannot be captured. (c) $A = 2.4503$, (d) $A = 2.4502$ to show the encroachment of chaos. (e) The LC circuits of two feedback oscillators coupled by a resistor. (f) The variation with time of the phase difference between the two oscillators when they are nearly locked, to show regular intermittency in contrast to (b).

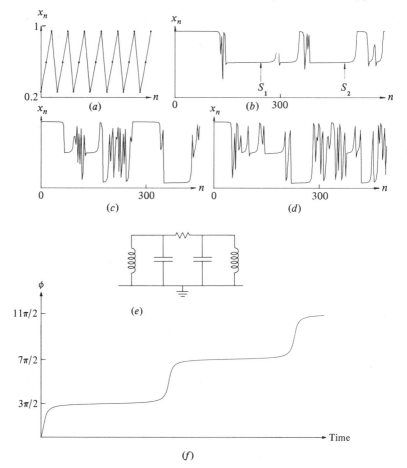

jumps occur at limit points, but there are alternative stable modes. And, with two-dimensional mapping (the impact oscillator) and mappings of higher dimensionality, the available space is so great that intermittency must be an unusual occurrence.

One last point – to concentrate on the 3-cycle iteration is to run the danger of obscuring the extraordinary complexity of the regime $A > 2.3022 \ldots$ after chaos has made its appearance. The chaotic state is interrupted, as A increases, by a bewildering variety of other periodicities. Fig. 4.20(b) shows three of period 5, indicated by E, an extra unattached 4-fold cycle at F, and four 6-fold cycles at G. Each of these will have its satellite bifurcations on the upper side and abrupt breakdown into intermittency and chaos on the lower side. And this is to enumerate only the simpler interruptions of the background chaos. For example, in the range of A between 2.355 and 2.356 there are over 50 extremely narrow stretches of 30-fold periodicity. It is expedient to beat a retreat at this point.

Successive approximation to the solution of equations[22]

The iteration we have been discussing is exactly the same as one elementary process of successive approximation used to compute solutions of the equation $x = f(x)$. One guesses a solution, x_0 say, and hopes to arrive at a better approximation $x_1 = f(x_0)$ by substituting the guess in $f(x)$; similarly $x_2 = f(x_1)$, and so on. It is now clear that convergence to the correct solution, $\bar{x}$, will only take place if the gradient $f'(\bar{x})$ lies between ± 1. Of course, when $f(x) = Ax(1 - x^2)$, successive approximation is unnecessary, but (4.9) is an example of an equation which it is tempting to treat in this way; if $A/Y_n > 1$, the curve represented by the right-hand side may have a gradient exceeding unity, so that this process of successive approximation is not guaranteed to converge.

On the other hand, the Newton–Raphson method is almost always rapidly convergent. To find the solution of $F(x) = 0$, we first find an approximate solution, perhaps from a rough graph, such as x_0 in fig. 4.23, and calculate the gradient at x_0 to determine where the tangent cuts the axis; this gives the next approximation, x_1, and repetition of the process soon leads to a good value for x. Formally, we write $x_{n+1} = x_n - F(x_n)/F'(x_n)$, reducing the successive approximation to the same method as before, but operating on $f(x)$ defined as $x - F(x)/F'(x)$. In the last stages of approximation, x will be so near $\bar{x}$ that the curve may be treated as a quadratic curve,

$$F(x) \sim \alpha\delta + \beta\delta^2, \quad \text{where } \delta = x - \bar{x}.$$

Then

$$F(x)/F'(x) = (\alpha\delta + \beta\delta^2)/(\alpha + 2\beta\delta)$$

and

$$\delta_{n+1} = \beta\delta_n^2/(\alpha + 2\beta\delta_n) \sim (\beta/\alpha)\delta_n^2.$$

Comparison with Ex. 4.8(b) shows that this is the equation for super-stability, so that extremely rapid convergence can be expected.

94   *Exercise 4.10.* Take an equation of the form (4.9), $x = 1.6 + 2 \sin x$, and try to find the solution, starting from $x_0 = \pi$. First put $f(x) = 1.6 + 2 \sin x$, and see what happens when you make successive approximations $x_{n+1} = 1.6 + 2 \sin x_n$. Then apply the Newton–Raphson method to $F(x) = 1.6 + 2 \sin x - x$, again starting from $x_0 = \pi$.

Two-variable mapping and bifurcations

At last we are ready to tackle the problem presented by the impact oscillator of fig. 4.9 when A is raised above A_{c0}, as given by (4.18). The representative point in fig. 4.12(d) leaves region C, crossing line (4) into the domain of instability. Within C the system has two convergent modes, both alternating in sign, but when it enters J one of them begins to diverge. If A is adjusted to a value near A_{c0}, and the system disturbed arbitrarily from its stationary state, the modes vibrate independently until one has died out and the other is very slowly decaying or growing according to the sign of $A - A_{c0}$. Then the Poincaré map of the behaviour

Fig. 4.23. The Newton–Raphson process of successive approximations.

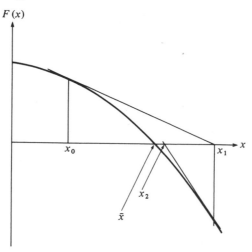

consists of an alternation between two points on the phase plane. There is no guarantee that when A is slightly greater than A_{c0} the slow growth of the vibration will be halted, but in this case it is. The full line in fig. 4.24 shows the resulting bifurcation for $\varepsilon = 0.8$, in which case $A_{c0} = 0.9878$. When $A > A_{c0}$ the forward and return journeys of the ball are no longer the same, with Y_0 lying alternately on the upper and the lower branch. There are two equivalent stationary states, mirror images, according to whether the larger Y_0 belongs to the forward or the return journey. If we allow that ε may not be the same for the two walls, these two states cease to be equivalent, and at the same time the bifurcation vanishes. One state, shown as α, has the ball moving faster to strike the softer wall, and slower to strike the harder wall; in β the pattern is reversed. The system falls naturally into α as A is increased from below A_{c0}, but β can be established at a higher value of A, only to collapse abruptly as A falls below A'. The true bifurcation is an accident of symmetry, when both ε are exactly the

Fig. 4.24. The first bifurcation of the impact oscillator. The full curve is drawn for $\varepsilon = 0.8$ on both walls, and when $A < A_{c0}$ the position and velocity of the ball are the same (apart from sign) when sampled in the forward and backward directions. Above A_{c0} the upper branch describes the forward velocity and the lower branch the backward, or vice versa. When ε is slightly different on the two walls, the broken line results. Even below A_{c0} the forward and backward velocities differ (α), and the effect is enhanced above A_{c0}; a new mode (β) enters at A', somewhat above A_{c0}. The innermost branches represent an unstable mode.

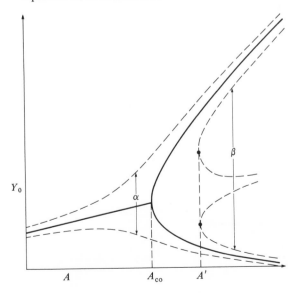

same. The higher bifurcations, to which we now turn, do not depend on this accident, and this first bifurcation should therefore be thought of as a separate phenomenon, not part of the sequence which really begins at the next bifurcation. There is no computational advantage to be gained by trying to collapse two iterations of (4.12)–(4.14) into a single formula; it is easier to continue sampling the motion every half-cycle, and to present every other result.

The development of the steady state as A is increased is illustrated in fig. 4.25, starting from A_{c0}. Up to A_{c1} ($= 1.344\,48$) the steady response has the same periodicity as the driver, so that a single point on the phase plane suffices to define it. The bifurcation at A_{c1} leads to a range of values in which alternate samples are different, while the next bifurcation at A_{c2} ($= 1.413\,35$) leads to a 4-fold period, represented by the crosses. It is a tedious business to find the points of bifurcation to sufficient accuracy with a desk computer. In contrast to the one-parameter system, there is no obvious way of finding the superstable points, even if they exist, but one may determine the relaxation time and how it diverges to infinity as the bifurcation is approached from below. Table 4.3 was constructed in this way; A_{cn} means the value of A at which bifurcation changes the period from 2^{n-1} to 2^n. The limiting value of the ratio of successive ΔA is clearly very close to Feigenbaum's value, $4.669\,2\ldots$. The theory developed for

Fig. 4.25. The next two bifurcations above A_{c0} with $\varepsilon = 0.8$, presented as stationary states on the phase plane for various values of A. $A_{c1} = 1.344\,48$, $A_{c2} = 1.413\,35$.

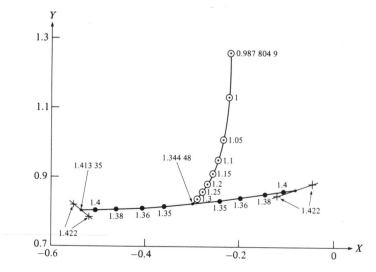

one-parameter mapping appears to hold equally well for two-parameter mapping, and not only in this particular case.

A plausible explanation of this similarity emerges when one examines the set of points (X, Y) making up the repeat pattern when n is large. Fig. 4.26 shows the 32 points (not all resolved) representing the steady state when $A = 1.4283$, more or less midway between A_{c5} and A_{c6}. The fact that they lie very close to a smooth curve strongly suggests that a single

Table 4.3

		$\Delta A_c \times 10^4$	ratio of successive ΔA_c
$A_{c1} = 1.344\,48$			
	>	688.7	
$A_{c2} = 1.413\,347$			5.786
	>	119.03	
$A_{c3} = 1.425\,249\,8$			4.806
	>	24.767	
$A_{c4} = 1.427\,726\,5$			4.660
	>	5.315	
$A_{c5} = 1.428\,258\,04$			4.663
	>	1.139\,8	
$A_{c6} = 1.428\,372\,02$			

Fig. 4.26. The cycle of 32 points when $A = 1.4283$, between A_{c5} and A_{c6}; not all are resolved.

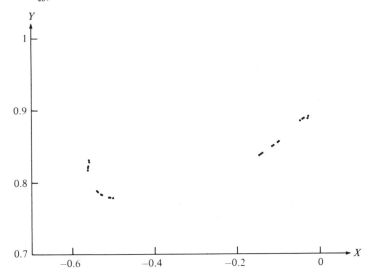

parameter, measuring distance along the curve, is enough to define each point, and thus to reduce the problem to a one-parameter mapping. Of course, this is not a rigorous argument, but it suggests the direction in which a rigorous explanation should be sought.

Having said this, one must point out that when A is further increased, to make the behaviour chaotic, the idea of a single curve on which all points will lie begins to look dubious, as fig. 4.27 shows. To be sure, the points are still confined to something like a one-dimensional domain, but the curve is folded many times upon itself between its ends, P and Q, with cusps at C and C', and hints of substructure, especially near C'. We must therefore not take for granted that everything is explained. Indeed, there is clearly much scope for mathematical refinement in the whole field of bifurcation and chaos. Similar plots of other chaotic two-dimensional systems,[23] extending to many hundreds of thousands of points, suggest that the substructure continues to get more complicated the further one goes, probably without limit. Thus ultimately the points do not fill the narrow bands indicated by those in fig. 4.27, but lie on discernibly separate lines whose number is unlimited. If the points were on one line, it would be a one-dimensional system; if they filled the band randomly it would be two-dimensional. The intermediate behaviour is said to have fractal[24] dimensionality between 1 and 2.

Fig. 4.27. The beginning of the pattern for the restricted chaotic state when $A = 1.435$. The more iterations, the finer the structure that appears.

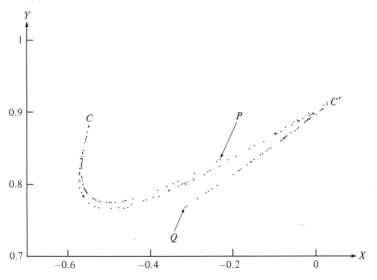

The region of phase space displayed in fig. 4.27 is quite small, and you may ask what happens if the system is started from a point somewhat distant from the locus traversed by the chaotic state. The answer is that unless one starts quite close to the locus the system is not drawn into it, but rattles around in phase space in the way we discussed earlier. Moreover, if   99
A is increased only a little above the value used in drawing the figure, the ends of the locus, *P* and *Q*, rapidly extend and become points of departure; sooner or later the representative point ceases to jump around on the locus of what we may call *restricted* chaos and moves off into the *utter chaos* which apparently involves the whole phase plane. The quasi-one-dimensional behaviour breaks down at this point, and we shall not attempt to follow the argument further. It is only to be expected that there should be more variety in a two-parameter than in a one-parameter system. A single example must suffice. In the range $1.282 < A < 1.293$ there is, as well as the simple unbifurcated steady state, a 3-fold steady state which collapses into the simple state at both ends of the range, without any preliminary bifurcation, presumably by limit-point instability, and without intermittency. Doubtless a diligent search would reveal many more intrusions into the pattern we have chosen to discuss in detail. But the search in two dimensions is considerably more tedious than in one. What the model has revealed is that there may be a great many effectively disconnected regimes, each developing with change of *A* in one of two ways: by bifurcations to restricted chaos, and thence to utter chaos; or by abrupt collapse into either an ordered or a chaotic state. The former route is taken by some of the ordered states (the principal mode of response which we have studied in detail, as well as the subharmonics) as *A* is increased. The latter route is taken as *A* is decreased and also by the 3-fold intrusive state as *A* is increased. Whether this comprises all possible forms of behaviour remains to be seen.

Examples of bifurcation and chaos

The investigation of bifurcations and chaotic states has only rather recently caught the imagination of research workers in physics, mainly because of a short, but eminently readable, review article by May,[17] who presented the issues involved in one-parameter mapping. It is still not clear whether the chain of bifurcations that he described and that we have illustrated here is a common phenomenon in real systems, or whether it is to be found mainly in simple models. A few such models, of which details will be found in the references, are:

1. The heavily damped non-linear vibrator (Duffing oscillator), in   86

which the potential rises more slowly than quadratic at large amplitudes, so that the natural frequency falls with increasing amplitude.[25] If the damping is low, the description of mode-jumping at the beginning of this chapter is probably valid; when the system is excited by a driving force at   78 its low-amplitude resonant frequency, and the driving frequency is then lowered, the amplitude of response rises in similar fashion to fig. 4.4, only to collapse abruptly when the drive can no longer supply the dissipated energy. With stronger damping, and correspondingly stronger drive, however, the response shows a chain of bifurcations. At the first bifurcation, alternate peaks begin to have different amplitudes, and after that the response continues to follow the driving force in general terms, but each of 2^n successive cycles is slightly different in form and amplitude. The Feigenbaum pattern is followed up to the point of chaos, and the chaotic state may continue for a while as the frequency is lowered further, before the moment of collapse.

2. Closely related to this is the driven simple pendulum, when the amplitude is forced to the point that the pendulum may go right over the top. The natural frequency falls with increasing amplitude as in the last example. The equation of motion for the angular displacement ϕ has the form, in appropriate units:

$$\ddot{\phi} + \lambda\dot{\phi} + \sin\phi = A\cos\omega t.$$

The same equation is obeyed, at least to a good approximation, by the Josephson junction, consisting of two superconducting films separated by a very thin oxide layer.[13],[26] It has been conjectured that the electrical noise generated by these junctions may have its origin in chaotic behaviour. Be that as it may, we have here an extension from simple demonstration models to a system which has many important applications as a measuring instrument of very high sensitivity.

3. A rather different example of bifurcation is provided by the circuit shown in fig. 4.28(a) or a variant, as in fig. 4.28(b).[27] If the diode rectifier behaved ideally, allowing easy passage of current in one sense and shutting off completely as soon as the polarity was reversed, the behaviour would be straightforward and would scale with the amplitude of the driving signal. A little time is needed, however, after reversal to eliminate the charge carriers present in the forward-biased diode; and the time depends on how large a forward current was carried. A large current in one cycle tends therefore to suppress the current in the next, and a mechanism is available for alternating response in successive cycles. In fact, a chain of bifurcations can be observed, as shown in fig. 4.28.

Fig. 4.28. (*a*) A resonant circuit including an imperfect diode, as described in the text; (*b*) a realization of the same behaviour under more controlled conditions. With the circuit parameters shown, a sinusoidal input at 50 kHz was suitable. The lower trace in (*c*)–(*g*) shows this input. As the amplitude is increased, the response measured at 0 initially has the same period, then is increased to twice (*d*), four times (*e*), eight times (*f*), shortly afterwards becoming chaotic (*g*), as shown by the structure inside each peak.

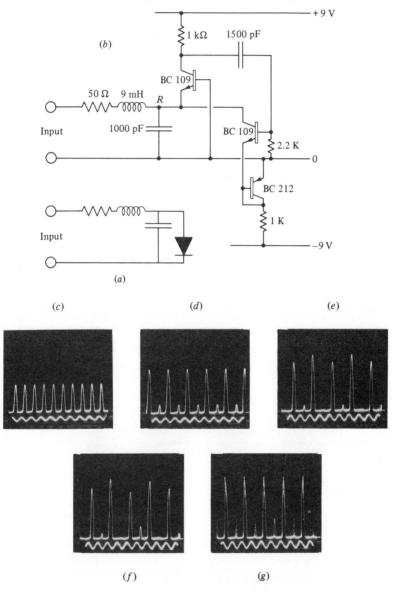

Computer simulations suggest that the Feigenbaum scheme is followed. Measurements on the circuit of fig. 4.28(c) could go only to the third bifurcation, insufficient to see anything more than that there is no disagreement with the computer result.

The discovery which, above all, has excited the interest of physicists is that under the right conditions the Feigenbaum scheme appears in a hydrodynamical context, thus providing hope of insight into the very real state of chaos represented by turbulence. That hope, to be sure, is unlikely to be realized if one is so optimistic as to expect a general solution to the problems, but to be able to describe the mechanism in even one case is something to be thankful for. The case in question is thermal convection, which has been studied ever since the time of Bénard[28] and Rayleigh.[29] In order to appreciate the main features of natural convection (i.e. not involving forced circulation), it helps to start with a simplified model resembling a domestic hot-water circulation system, but (like some earlier installations) without a pump.

In fig. 4.29, the upper horizontal tube is maintained at a constant temperature, labelled $\theta = 0$, while the lower horizontal tube is heated to θ_0; water or some other fluid circulates at will. A short qualitative argument suffices to show that there is a critical value of θ_0 needed to maintain circulation. The Exercise that follows invites a fuller working out. If there is no circulation, thermal conduction through the water or the tube walls leads to a linear temperature distribution in the vertical tubes, the same in both, and there is nothing to disturb the hydrostatic equilibrium. If circulation starts, however, with mean velocity $\bar{v}$ in the sense shown (both senses are of course equivalent) the upward flow of hot water on the left raises the mean temperature in this limb, while the downward flow of cold

Fig. 4.29. Diagram of a convective circuit, to show notation.

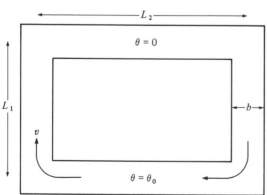

water lowers the mean temperature in the right-hand limb. Consequently the mean density is less on the left than on the right, and hydrostatic forces tend to maintain the circulation. If the mean temperature difference between the vertical limbs is ΔT, the pressure available to drive the circulation is $g\rho' L_1 \Delta T$, where $\rho' = -\mathrm{d}\rho/\mathrm{d}T$, the temperature coefficient of density ($\rho' = \rho\beta$, where β is the volume expansion coefficient). Now for slow circulation rates, $\Delta T \propto \bar{v}$, as may be understood by noting that the sign of ΔT reverses with that of $\bar{v}$. When $\bar{v}$ is small, then, the driving pressure $\Delta p \propto \bar{v}$; when $\bar{v}$ is large, however, the water passes through the horizontal limbs so quickly that it has no time to take up the temperature 0 or θ_0, and the driving pressure begins to fall, as shown schematically by curve 1 in fig. 4.30(a).

On the same diagram we now show how the actual mean velocity $\bar{v}$ depends on Δp, being limited by viscosity. So long as the flow is laminar, $\bar{v} \propto \Delta p$, as shown by curve 2. If the curves intersect, flow is possible and the steady velocity is given by the point of intersection, V. It must be remembered that the scale of curve 1, but not that of curve 2, is affected by the temperature difference, the vertical height being proportional to θ_0. If θ_0 is too small, there is no intersection, and clearly the critical condition for the liquid to become unstable and begin to circulate is that the initial linear portion of curve 1 should have the same slope as curve 2. It only requires a slight excess of θ_0 over the critical value for V to rise to a significant magnitude, as indicated by fig. 4.30(b), which is constructed from the sketched form of curve 1.

Having established the existence of a critical value of θ_0, let us now work out the details. For simplicity, and because it will tie in with the next stage

Fig. 4.30. (a) Curve 1 indicates how the driving pressure in the convective circuit rises linearly to a maximum and then falls; the line 2 is the driving pressure needed to maintain a mean velocity $\bar{v}$. If there is an intersection V, convective flow occurs. (b) Schematic diagram showing convective flow starting, in either direction, at a critical value of θ_0.

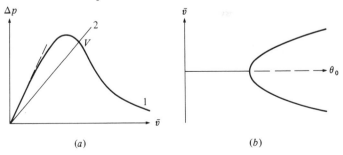

(a) (b)

of the discussion, we choose a quasi-two-dimensional model, in effect assuming the tubes to be made of flat sheets and to be very wide normal to the paper. The separation of the sheets is b.

Exercise 4.11. Demonstrate the following propositions, assuming steady flow:

(a) If no heat crosses the walls of the vertical limbs, conservation of heat energy demands that the temperature distribution in the vertical limbs, $\theta(y)$, obeys the equation

$$C\bar{v}\theta - \kappa\, d\theta/dy = \text{constant},$$

in which C is the thermal capacity per unit volume of the fluid, and κ its thermal conductivity; both are assumed temperature-independent, and y increases in the direction of v.

(b) The solution of this equation which satisfies the temperature constraints at top and bottom, $\theta(0) = 0$ and $\theta(L_1) = \theta_0$, is

$$\theta = \theta_0(e^{\alpha L_1} - e^{\alpha y})/(e^{\alpha L_1} - 1), \quad \text{where } \alpha = C\bar{v}/\kappa.$$

When αL_1 is small,

$$\theta \sim \theta_0[1 - (2y + \alpha y^2)/(2L_1 + \alpha L_1^2)],$$

and hence, on averaging over the length of the limb,

$$\bar{\theta} \sim \tfrac{1}{2}\theta_0(1 + \alpha L_1/6).$$

Therefore

$$\Delta T = \theta_0 C\bar{v}L_1/6\kappa,$$

and

$$\Delta p = g\rho' L_1\, \Delta T = g\rho' C\theta_0\bar{v}L_1^2/6\kappa.$$

(c) Now consider laminar viscous flow between two parallel planes at $x = \pm\tfrac{1}{2}b$. The velocity distribution $v = v_0(1 - 4x^2/b^2)$ is compatible with a pressure gradient $dp/dy = 4\eta v_0/b^2$. Writing $\bar{v} = \tfrac{2}{3}v_0$ and $dp/dy = \Delta p/2(L_1 + L_2)$, we have

$$\Delta p = 12\eta\bar{v}(L_1 + L_2)/b^2.$$

(d) The critical condition, $\theta_0 = \theta_{0c}$, is that which makes the expressions for Δp in (b) and (c) identical:

$$g\rho' C\theta_{0c}L_1^2 b^2/\kappa\eta(L_1 + L_2) = 72.$$

The quantity $L_1(L_1 + L_2)/b^2 = S$ is a dimensionless number describing the shape, and $g\rho' C\theta_0 L_1^3/\kappa\eta$ is another dimensionless number, the *Rayleigh number*, Ra. The condition for steady circulation is

$$\text{Ra} > \text{Ra}_c; \quad \text{Ra}_c = 72. \tag{4.24}$$

Exercise 4.12. Bearing in mind that the maximum temperature difference between rising and descending flows is θ_0, show that the mean velocity of circulation cannot exceed $g\rho'\theta_0 L_1^2/12\eta S$.

Since we cannot expect the steady state to be reached until some fluid at least has completely traversed the vertical limbs, we may take $12\eta S/g\rho' L_1 \theta_0$ as a measure of the time needed to establish the steady state. In practice a much larger time than this is needed for precise measurements. To achieve rapid equilibrium, η/ρ' must be small. For water at 20 °C, $\eta/\rho' \sim 5 \times 10^{-3} \text{ m}^2 \text{ Ks}^{-1}$, while for liquid helium, which at 4.2 K has very low viscosity and high expansion coefficient, it is 1.5×10^{-7}, smaller by a factor of 3×10^4. It is the great rapidity of the processes in liquid helium, together with the possibility of high sensitivity in temperature measurement, that makes it an attractive medium for studying convection.[30]

The example we have just examined is not far removed from the more interesting case of free convection in a fluid above a heated plate. Quite frequently the initial circulatory motion, when the temperature gradient just exceeds the critical value, takes the form of long rolls, as sketched in fig. 4.31, the precise configuration depending on the shape of the container. It is not surprising to find that the critical condition takes the same form as (4.24), with circulation starting when the Rayleigh number exceeds a certain value. If the fluid is of infinite extent horizontally, so that there is nothing to prevent the rolls assuming the optimal shape, Ra_c is calculated to be 1708,[31] and this accords well with experimental determinations. The value $S = 24$ needed in (4.24) may seem a little high if one makes reasonable guesses of L_1, L_2 and b appropriate to the flow pattern in fig. 4.31. It should be noted, however, that in fig. 4.31, in contrast

Fig. 4.31. The convection cell used by Libchaber and Maurer; the length, normal to the paper, was 1.5 mm. The black circles represent two very small thermometers embedded in the top surface, in thermal contact with the convecting fluid, liquid helium at about 4 K. Both top and bottom faces are of highly conducting metal, while the side walls match the conductivity of helium. The whole apparatus is suspended in a vacuum.

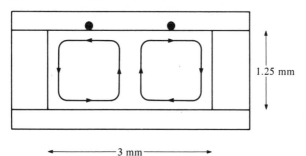

1.25 mm

3 mm

to fig. 4.29, heat can flow sideways as well as vertically to reduce the temperature differences, and therefore the driving force available to maintain circulation. This will put up the value of Ra_c.

As the Rayleigh number is increased above Ra_c, the pattern of circulation undergoes a considerable number of transitions, not all of which are understood. Direct observation of the flow is normally out of the question, but bolometers embedded in the lower surface of the top plate gives significant information on the behaviour of the liquid, without revealing the whole picture. Here we shall note only one part of the story, as described by Libchaber and Maurer.[32] When Ra/Ra_c reaches about 8, the bolometers begin to record regular oscillations at about 0.3 Hz, caused, there is good reason to believe, by a rippling motion of the division between the two rolls. This presages an extended set of transitions to a state in which the motion of this division is chaotic. Passing over one or two complexities on the way, we come to the point where $Ra/Ra_c = 41$ and the fundamental frequency, which is now 0.45 Hz, is accompanied by a subharmonic at half the frequency. This is exactly what one expects to appear at a period-doubling bifurcation as in the circuit discussed above (fig. 4.28). As Ra/Ra_c proceeds from 42.7 to 43.0 more bifurcations occur at successively closer intervals; Libchaber and Maurer were able to observe 4-fold, 8-fold and 16-fold patterns before the behaviour became aperiodic.

It is not to be expected that the experimental determination of the bifurcation points will approach the accuracy of the computations discussed earlier, and without this sort of accuracy there is no hope of confirming that they follow the Feigenbaum scheme. The most that can be said is that the bifurcations get rapidly closer, and that the ratio of successive gaps is 3.5 ± 1.5, not in disagreement with Feigenbaum's $4.669 \ldots$.

It has fairly recently been found that the hydraulic circuit as in fig. 4.29 may also show bifurcations leading to chaos.[33] One must not, however, jump to the conclusion that the route from regular, streamline flow to turbulence is always by way of bifurcations. On the contrary, natural convection seems to be a very exceptional example. Nevertheless, all processes in which streamline flow gives way to turbulence have this in common, that the fluid flow follows a well-defined differential equation, the Navier–Stokes equation. Over short time intervals the local development of the flow pattern is clearly smooth, and, in principle, predictable; but, as in the simple examples of chaos we examined above, the long-term behaviour is far too sensitive to tiny changes of the initial pattern to allow detailed predictions for anything beyond the immediate future.

The prevalence of chaos

[When one finds simple mathematical operations, not pathological in themselves but arising from observation, yielding a rich, even
grotesque, harvest of bifurcations and chaos, one feels the ground shifting
underfoot. Can it be that the systematic reduction of complex processes to
their basic constituents, obeying laws of marvellous simplicity, has left us
with a body of knowledge whose usefulness is rather problematical? It has
been the habitual claim of physicists that they could make predictions
whose verification underpinned the laws and conferred on science a
validity that no other branch of learning could aspire to. Was this a
delusion? Of course not, but the claim may have been overoptimistically
expressed.

It was always recognized that complexity might preclude detailed
prediction – no one ever hoped to follow the motion of each molecule in a
gas. But long before statistical mechanics provided a theoretical foundation, it was clear that the average properties of pressure, velocity,
temperature, etc., obeyed quite straightforward laws of thermodynamics
and hydrodynamics. Straightforward though they were, the equations
expressing them were still capable of yielding highly irregular solutions,
and this time there is no molecular complexity to explain turbulence away
– it is intrinsic to the equations. This should have been enough to alert us
to the potential in almost any non-linear differential equation to surprise
us by the diversity of its solutions.*

Nevertheless, out of turbulence and chaos it may still be possible to
extract some useful averages. For example, no motion of the atmosphere is
free from turbulence (look at the leaves on a tree, or the fluttering of a flag,
even on a still day), yet short-term weather forecasting is on the whole very
successful. The large air movements cannot be abruptly altered, but have a
natural time-scale of many hours, or even days (more correctly, a wide
spectrum of time-scales, but still hours or days for the processes affecting
the weather). The only processes which are predictable over weeks, if
indeed there are any, are of such a large size that they play next to no part
in the details of the weather. There is little hope of ever knowing whether
this day fortnight will be wet or dry – only that some weather patterns are
more likely than others. To say this is only to reflect what is intrinsic in fig.

---

* It should not be forgotten that the mapping procedures discussed in this
chapter arise out of the way we have chosen to deal with non-linear
differential equations. The result looks rather simple, but this is an illusion. If
you are not convinced, try to devise a differential equation for $x(t)$ which,
when sampled at regular intervals, yields (4.22).

4.19: chaotic systems may be predictable for a short while ahead, but the inevitable small perturbations will grow to dominate their environment; and there seems to be nothing one can do about it.

It is tempting to see the life of an extended society as a species of chaos – normally it is restricted chaos in that the observed actions constitute only a small fraction of the possibilities. The analogy becomes especially persuasive when one recalls that ancient commonplace that no action, however small, but has consequences that spread through the community: 'for want of a nail the shoe was lost, for want of a shoe . . .'.[34] The implication is that one must not expect the theories of social science to be predictive except for short times ahead, or for certain processes which can develop only slowly, like the fundamental principles of a political party. It may be, for instance, that grand overall economic theories can be devised, but that the limited economic theories that governments long for will never have predictive validity for far enough ahead to be any use. Be that as it may, it is desirable that those social scientists who seem to strive to make their discipline conform as closely as possible to the perceived ideal of physics, should recognize that they may be imitating physical procedures at the very point where they are least reliable. And for their part physicists should not allow themselves to be gratified overmuch when philosophers of science select physics as the typical science. The phenomenon of chaos is a salutory reminder of the frailty of human endeavour, and it may be that the recognition of the limitations of mathematical prediction will prove the most typically scientific aspect of physics.]

5

Elementary types of catastrophe

In the first three chapters we examined the stability and the response of systems which could be treated as linear. Here, and in the non-linear systems which were the concern of chapter 4, we met from time to time examples of sudden changes which might be massive, or a bifurcation at which alternative paths appeared. It is now time to address the problem of the behaviour of systems as they pass through such critical points. What happens after that? The question has no universal answer; for example, until the critical Euler load is reached, the behaviour of a thin strut can be analysed on the basis of very little detailed knowledge of its properties; but once the critical load is exceeded it may snap (if brittle) or sink slowly to the floor (if plastic) or, if it has a reasonable range of elasticity, remain in a bent position indefinitely. No general theory will exempt the physicist who needs to understand these things from undertaking a serious study of materials science. Similarly, the equilibrium shape of a pendent drop growing on the end of a pipette may be calculated from hydrostatic principles with the surface tension as an extra parameter. So long as the drop is able to support itself, the same theory applies to all liquids; beyond that the viscosity will determine whether it falls cleanly or trails a fine thread behind it – and if non-Newtonian fluids are considered a still wider field of possibilities is opened up. Each case like this needs to be studied in its own terms, and I shall make no attempt to draw up categories. But there still remains an extensive residuum of systems that pass through their critical points without requiring a new specification of their properties. Struts which behave elastically beyond the critical Euler load all behave in the same way, whether of steel or nylon. Moreover, we shall find that the modes in which such respectable systems can develop are very strictly limited, with the result that they can be characterized by a few parameters.

Let us recall some systems we met earlier which exemplify the different types, and then proceed to a catalogue of examples to show the wide applicability of the analysis.

> *Exercise 5.1.* Take another look at the wheel and pulley of Ex. 1.4, 4
> and show that, when the wheel is turned through θ from the
> position where the rim load is at the bottom, the potential energy
> takes the form
>
> $$V = C - A\theta - B \cos \theta,$$

where $A \propto m$ and B is determined by the pulley, and assumed constant. This function is shown in fig. 5.1 for various values of A. The critical value of A, $A_c = B$, is the largest for which the curve has minima, and for which positions of stable equilibrium exist. Write θ as $\pi/2 + \varepsilon$ and expand $\cos \theta$ as a Taylor series about $\pi/2$ to show that, apart from constants, for each value of A

$$V \sim a\varepsilon - b\varepsilon^3, \quad \text{where} \quad a = B - A \quad \text{and} \quad b = \tfrac{1}{6}B.$$

When $a > 0$, i.e. $A < A_c$, V has a minimum at $\varepsilon = -(a/3b)^{\frac{1}{2}}$; around this minimum

$$V \sim (3ab)^{\frac{1}{2}}\varepsilon^2 - b\varepsilon^3.$$

Fig. 5.1. Behaviour of the potential in Ex. 5.1, showing positions of stable (S) and unstable (U) equilibrium, which converge at the cross-lines on the curve for $A = B$. The value of C is chosen to display the curves best, and has no physical significance. The numbers attached to the curves indicate the values of A/B.

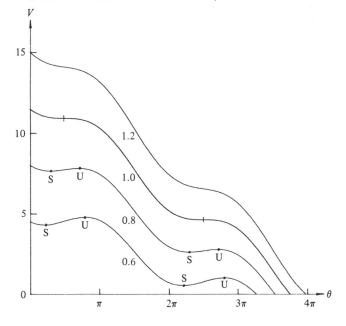

The first term governs the frequency of small oscillations about equilibrium, and $\omega_0 \propto a^{\frac{1}{4}}$ when a is small.

Exercise 5.2. Now return to the Euler strut to see how it behaves after the critical load is exceeded, on the assumption that it remains perfectly elastic. As in fig. 5.2, the shape of the strut may be defined by the inclination to the vertical of the tangent at a distance s from the top, measured along the strut. Show that the curvature is $d\phi/ds$ so that, without approximation,

$$Wy - EI \, d\phi/ds = 0.$$

Hence, noting that $dy/ds = -\sin \phi$, derive the differential equation for the shape,

$$d^2\phi/ds^2 + \alpha^2 \sin \phi = 0, \quad \text{where } \alpha^2 = W/EI.$$

This has the same form as the equation of motion of a simple pendulum when its amplitude is not small,

$$d^2\theta/dt^2 + \omega_0^2 \sin \theta = 0.$$

By comparing these two equations, and remembering that the pendulum has a lower frequency at larger amplitudes, show that the strut deflects, but finds new equilibrium positions as W is raised above the critical value, W_c. For small excess weight, the deflection is proportional to $(W - W_c)^{\frac{1}{2}}$. In fact, if it is assumed that the strut is perfectly elastic there is an equilibrium shape for every

Fig. 5.2. Notation for large-displacement theory of the Euler strut.

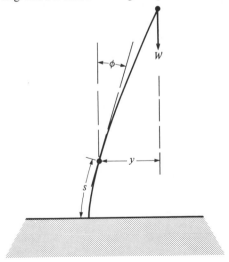

value of W, and the shape is progressively modified, without any discontinuity, until when W is large the load hangs down below the base. The solution of this problem (the *elastica*) is a standard exercise in applied mathematics.[1]

Exercise 5.3. The plank of Ex. 1.6 does not behave in the same way. If it is tilted through θ without slipping, show that the centroid C is lowered by an amount $(R+H)(1-\cos\theta) - R\theta\sin\theta$, i.e.

$$\tfrac{1}{2}(H-R)\theta^2 + \tfrac{1}{24}(3R-H)\theta^4 + \cdots.$$

When $H=R$ the quadratic term is zero, and the positive coefficient of the quartic term shows that the plank will continue to tip, once tilted from the level.

Classification

These three exercises exemplify the three types of behaviour of a system described by one variable, as the change of parameters takes it from a stable to an unstable state.* In each case the potential energy contains a quadratic leading term which is positive for stable equilibrium and negative for unstable. The next higher term in the series expansion of V determines what happens next:

Limit point (or asymmetric) instability. $V = -\alpha x + \beta x^3$, with $\beta > 0$, as in Ex. 5.1. When $\alpha = 0$ the system is stable against negative displacement, unstable against positive. When $\alpha > 0$ there is a single stable position, at $(\alpha/3\beta)^{\frac{1}{2}}$.

Stable symmetric transition. $V = -ax^2 + cx^4$, with c positive, as in Ex. 5.2. When $a \leqslant 0$ the system is stable for either sign of displacement, and when $a > 0$ takes up either of two stable positions where $dV/dx = 0$, at $\pm(a/2c)^{\frac{1}{2}}$.

Unstable symmetric transition. $V = -ax^2 + cx^4$, with c negative, as in Ex. 5.3. When $a < 0$ there is a metastable position of equilibrium at the origin, and when $a > 0$ there is no stable position anywhere.

These are illustrated in fig. 5.3. Clearly one can continue the series expansions (if they are convergent) to higher terms, but we have already obtained the most important results. If we have to extend the analysis beyond this point, we can deal with each problem as it arises, by whatever method seems appropriate. An important difference between the first case

---

* One may occasionally meet pathological cases, such as would occur in Ex. 5.3 if the coefficient of the quartic term vanished, leaving the coefficient of θ^6 as the arbiter of stability. But these are rare and I shall ignore them.

and the next two is that, in the last two, when $a > 0$ the symmetrical configuration is still an equilibrium state, albeit one of unstable equilibrium. In the first case (limit point) there is no equilibrium state, stable or unstable, when $\alpha < 0$. It may also be noted that the symmetric transitions involve a bifurcation, with one stable solution splitting into two equivalent solutions; by contrast, at a limit point the only solution simply vanishes. Recognition of these differences may help us to assign any particular critical transition to its appropriate category.

Fig. 5.3. The three basic transitions: (*a*) the limit point instability, $V = -\alpha x + \beta x^3$, with α positive in (1), zero in (2) and negative in (3); (*b*) positions of stable (full line) and unstable (broken line) equilibrium for case (*a*), showing convergence at the limit point, $\alpha = 0$ (cf. fig. 5.1); (*c*) the stable symmetric transition, $V = -ax^2 + cx^4$, with c positive and a positive in (1), zero in (2) and negative in (3). The curves are shifted vertically for clarity; (*d*) equilibrium positions as in (*b*); (*e*) the unstable symmetric transition, $V = -ax^2 + cx^4$; $c < 0$, with notation as in (*c*).

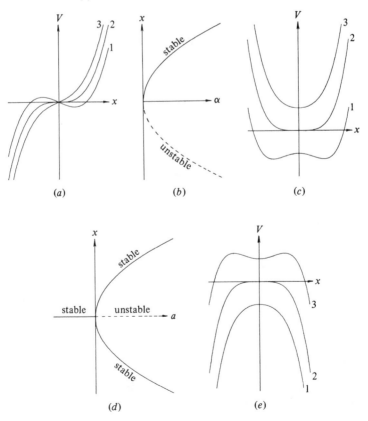

Let us now consider a number of examples of each type of instability, to appreciate how frequently they arise in almost every branch of physics. Do not, however, expect many examples from modern physics. It is not that they are any less frequent, but that it would take too long to explain the background to, for example, stellar instability, or nuclear fission, or the many plasma instabilities that bedevil attempts at nuclear fusion.

As you go through these exercises take note that only stable symmetric transitions are reversible. An elastic strut, loaded beyond its critical load, will return smoothly to the vertical position as the load is reduced. But the systems of Ex. 5.1 and 5.3, once taken beyond their critical points, are beyond recall by a trivial reversal of the process that took them over the brink. Limit point instabilities and unstable symmetric transitions are either wholly irreversible or exhibit hysteresis, requiring a sizeable change of parameter to recover their original form. Thus the soap film in Ex. 5.6 and the tube of Ex. 5.15 are lost beyond recall, while the spring in Ex. 5.17 is hysteretic.

Limit point instabilities

Exercise 5.4. Krishnan and Banerjee[2] developed a method for measuring the anistropy in the magnetic susceptibility χ of simple crystals that could only be prepared in very small amounts (< 1 mg). They suspended the crystal from a fine silica torsion fibre between the poles of an electromagnet giving a uniform horizontal field B. The torsion head was adjusted so that the crystal did not turn as B was increased; this means that one magnetic axis is parallel to B. If this is the axis along which χ is greatest, the crystal will now try to stay parallel to B as the torsion head is turned, and will turn more slowly than the head. Indeed, if the torsion fibre is very thin the crystal will hardly turn at all. Eventually, however, as the turning continues, a point will be reached when the magnetic torque on the crystal can hold it no longer, and the crystal will suddenly start to spin. The amount the torsion head must be turned to reach the point of instability enables the anisotropy of χ to be calculated.

If the mass susceptibility of the crystal is χ_1 and χ_2 along the two principal directions at right angles ($\chi_1 > \chi_2 > 0$), show that a direct calculation of the magnetic moment, and hence the magnetic torque, when there is an angle θ between B and the axis of χ_1, is consistent with an expression for the potential energy of the crystal:

$$V_{\text{mag}} = -\tfrac{1}{4}m\mu_0 B^2(\chi_1 - \chi_2)\cos 2\theta.$$

Also if the torsion constant of the fibre is c and the torsion head has been turned through ϕ, the strain energy in the fibre is

$$V_s = \tfrac{1}{2}c(\phi - \theta)^2.$$

Show that the total energy, $V_{\text{mag}} + V_s$, has a minimum with respect to θ, at which the crystal is in stable equilibrium, only when $\theta < \pi/4$; and that the anisotropy of χ is related to ϕ_{c}, the value of ϕ at the point of instability, by

$$\tfrac{1}{2}m\mu_0 B^2(\chi_1 - \chi_2) = c(\phi_{\text{c}} - \pi/4).$$

The suggested solution of the problem reveals its similarity to Ex. 5.1. It would be at least as easy to determine directly the angle beyond which the torques due to the magnetic moment and the torsion wire could not balance.

Exercise 5.5. This problem is most easily tackled by calculating forces rather than energy. The upper plate of a parallel plate capacitor, fig. 5.4, is suspended from a spring so that when the capacitor is uncharged it hangs a little way above the lower plate. The upper plate is earthed; show that as the potential of the lower plate, V, is increased the upper plate is attracted downwards, and becomes unstable when the spacing is reduced to two-thirds of its original value. It is then attracted hard against the lower plate.

Exercise 5.6. (a) A soap film formed between the ends of two open circular tubes, as in fig. 5.5(a), narrows down so that at every

Fig. 5.4. Capacitor plate on spring, for Ex. 5.5.

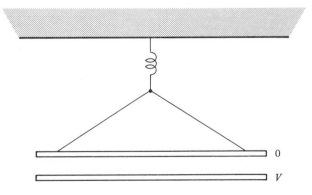

Fig. 5.5. (*a*) Soap film formed between the ends of two open tubes of internal diameter 26 mm. The blobs are soap solution which has drained off. (*b*) Theoretical shape of film, drawn to correspond to (*a*). If you trace this and compare it with (*a*) you will see the agreement is excellent. The marks on the curve denote the end of the right-hand tube. The curve terminates at the point where a film of that shape would be at its limit point. (*c*) $S/2\pi R^2$ plotted against b/R for different values of l/R as indicated on the curves. The middle curve is at the limit point.

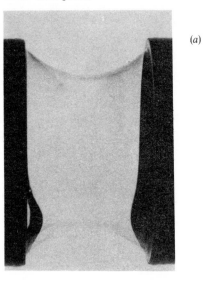

(*a*)

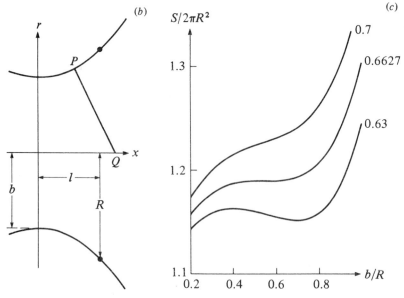

point the two principal curvatures are equal and opposite; this ensures that there is no pressure difference between inside and outside. The principal curvatures at P in fig. 5.5(b) are found by drawing a line through P normal to the film and sectioning the film by two planes containing this line – one lying in the paper and the other normal to it. The curvature in the former is given by the standard expression $(d^2r/dx^2)/[1+(dr/dx)^2]^{\frac{3}{2}}$. Show that the latter section has its centre on the axis at Q. Hence determine the curvature and write the differential equation for the film as

$$r\, d^2r/dx^2 = 1 + (dr/dx)^2.$$

The solution is $r = b \cosh (x/b)$, where b is shown in the diagram. Show that x/r has a maximum value of 0.6627 when $x/b = 1.200$. This means that for tubes of radius R, the film can be formed so long as the gap between the tubes is less than $1.3254R$; the radius of the neck cannot be less than $0.552R$. When the tubes are pulled further apart the neck collapses, leaving a flat film across the end of each tube.

(b) The relationship of this problem to the standard form of limit point instability is clear if you remember that the free energy of the film is proportional to its area, which is minimal in equilibrium. Try writing the shape of the film as $r = b \cosh (x/c)$, with $c = l/\cosh^{-1} (R/b)$; this gives a family of curves with different neck radii, b, but all fitting a pair of tubes of radius R, with a gap $2l$. Show that the area of the film is

$$S = 2\pi(R\lambda/\cosh^{-1} \rho)^2[\sinh^{-1} z + z(1+z^2)^{\frac{1}{2}}],$$

where $z = \lambda^{-1}(1-1/\rho^2)^{\frac{1}{2}} \cosh^{-1} \rho$, $\rho = R/b$ and $\lambda = l/R$ (this involves a certain amount of manipulation). As fig. 5.5(c) shows, S has a minimum only if $l < 0.6627R$.

Exercise 5.7. Here we shall examine the instability from the point of view of small vibrations about the position of equilibrium. The mass in fig. 5.6 is constrained to move in a vertical plane (the plane of the paper), being held up by two compression springs, each having spring constant μ, and unstrained length a_0. In equilibrium, clearly $mg = 2\mu(a_0 - a) \sin \phi$. For small displacements in the plane, m moves in a quadratic potential. If it is in stable equilibrium there are two directions at right angles along which it can vibrate in a straight line; these are the normal modes, whose frequencies in general are different. If the equilibrium is unstable, the frequency of at least one of the normal modes is imaginary. By symmetry, in this case, the normal mode vibrations are along a vertical and a horizontal line. Show that the frequency of vertical motion is given by

$$\omega_V^2 = (2\mu/m)[1 - (a_0/a)\cos^2\phi],$$

while that of horizontal motion is given by

$$\omega_H^2 = (2\mu/m)[1 - (a_0/a)\sin^2\phi].$$

If the separation of the feet happens to be just right so that on increasing m, ω_V becomes zero when $\phi = \pi/4$, then simultaneously ω_H also vanishes. If the separation is greater, increasing m causes ω_V to vanish first, and the mass falls straight down. This is a limit point instability. If, on the other hand, the separation is less, ω_H vanishes first and the system skews either to right or left. This is a symmetric instability, whose discussion we defer.

Considering the limit point only, suppose that l is adjusted to be pa_0. Show that the critical mass at which m collapses is given by

$$gm_c/2\mu a_0 = (1 - p^{\frac{2}{3}})^{\frac{3}{2}}.$$

Now let us see how ω_V varies with small changes of m. Some care is needed since a and ϕ change rapidly as m approaches m_c. Express $\sin\phi$ in terms of a, so that

$$mg = 2\mu(a_0 - a)(1 - p^2a_0^2/a^2)^{\frac{1}{2}}.$$

Fig. 5.6. Schematic (and idealized) arrangement to allow a mass to be supported on compression springs, as in Ex. 5.7. The springs are enclosed in frictionless telescopic tubes.

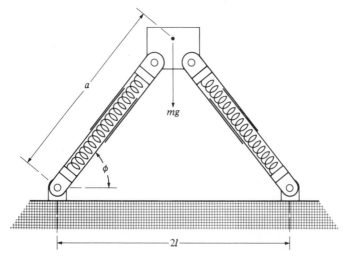

Then put m equal to $m_c(1 + \alpha)$, and a equal to $a_c(1 + \beta)$, and write this equation in the form of a series expansion in α and β. The term proportional to β vanishes, leaving as leading terms

$$\alpha = -\tfrac{3}{2}\beta^2 p^{\frac{3}{2}}/(1 - p^{\frac{3}{2}})^2.$$

Near the critical point, $a - a_c \propto (m_c - m)^{\frac{1}{2}}$. Thus, although ω_V^2 is a linear function of $(a - a_c)$, ω_V falls to zero as $(m_c - m)^{\frac{1}{4}}$, just as we have found for other limit point instabilities.

Something very similar to this simple model, though much harder to analyse, is the steel strip arch shown in fig. 5.7. Here also, when the feet of the arch are far apart, adding weight at the centre leads to a sudden symmetrical collapse ((a) and (b)). But with the feet closer together it slops over sideways at the first instability ((c) and (d)). This is another example of a stable symmetric transition such as we shall discuss in the next section.

A limit point instability is easily demonstrated with a pocket steel measuring tape, of the sort whose tape is bent in cross-section so that it stays stiff when unwound. If you lay such a tape flat on a table, with the concave side upwards, and then gently push the end over the edge, suddenly it will kink at the table-edge and collapse vertically to the floor. On pulling the tape back you will find very marked hysteresis – the instability has caused it to collapse to a quite different configuration, from which it cannot recover by a slight reversal of the overhang.

Try repeating the experiment with the steel tape, but this time turning it over so that it is convex. The critical point now comes much sooner, and, instead of collapsing, the tape kinks gently in such a way that its end begins to fall sideways rather than straight down. If the tape is free of defects it should with equal ease deflect itself to the right or the left, and in a reversible manner. This is yet another stable symmetric transition. On pushing the tape further out it will reach a limit point at which it will collapse irreversibly just as it did when placed the other way up.

Other examples of limit point instability are found in chapter 4, both in the cases of mode-jumping and in the loss of synchronism of the impact oscillator when the vibration amplitude is lowered. In all these instances the solution of the equations disappears at a certain critical value of the adjustable parameter (frequency or amplitude), just as in the above examples the instability occurs when the equation defining the point of stable equilibrium ceases to have a solution.

Exercise 5.8. This example is slightly different, in that the system is autonomous, i.e. self-driven; without any external aid it runs towards a limit point. Consider the circuit of fig. 2.3, with the

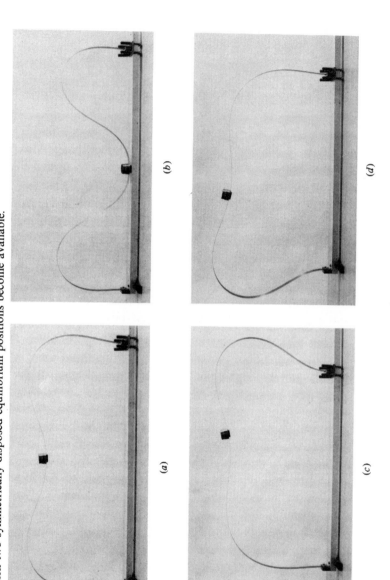

Fig. 5.7. An arch of spring-steel strip, about 25 × 0.5 mm cross-section and 1.7 mm long. The pot in the middle can be filled with lead shot until a transition occurs, and its position on the strip can be adjusted to achieve symmetrical behaviour. In (a) and (b) the feet of the arch are quite far apart, and a limit point instability occurs; in (c) and (d) the feet are closer and the first instability is a stable symmetric transition in which two symmetrically disposed equilibrium positions become available.

(a)

(b)

(c)

(d)

inductor removed. So long as the current through R and C is less than the saturation current of the amplifier, the total resistance is negative, but, as the current increases, saturation causes the resistance to become positive. A typical $i-V$ curve for the arrangement is shown in fig. 5.8(a).

Show that the current through C obeys the equation

$$R'\,di/dt + i/C = 0,$$

in which R' is the local slope of the $i-V$ curve, dV/di. Hence show that any positive i tends towards the value at the extremum P, while any negative i tends towards the value at P'. Also, as i approaches either point, $|di/dt| \to \infty$.

To understand what happens after P or P' is reached, note that at Q the voltage across the capacitor, and hence the charge, is the same as at P. The current is different, but in the absence of an inductor there is nothing to resist a change of i. The state of the circuit can thus jump from P to Q at a speed limited only by residual inductive effects in the circuit or the amplifier. The waveform corresponds to the process $PQP'Q'P$, and is shown in fig. 5.8(b).[3]

Stable symmetric transitions

Exercise 5.9. Two equal small bar magnets are attached to a piece of fine thread so that each hangs horizontally, one above the

Fig. 5.8. (a) $i-V$ diagram for a negative-resistance element, showing the path taken by multivibrator oscillations, which lead to the $i(t)$ behaviour shown in (b).

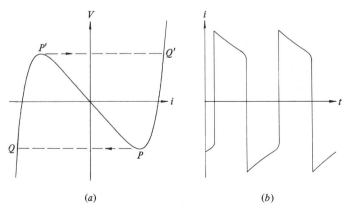

(a) (b)

other, and is free to turn about a vertical axis.* In the absence of an externally applied magnetic field, they set themselves anti-parallel, but when a weak uniform horizontal field is applied and slowly increased, they adjust themselves in a compromise configuration, balancing the tendency to be mutually antiparallel with the tendency for both to lie along the field. If each magnet produces a field b acting on the other, show that in an external field B, less than $2b$, there is an angle 2θ between them, where $\cos \theta = B/2b$. This is a typical bifurcation, like the Euler strut; as long as $B > 2b$ both magnets point exactly along B, but as B is reduced past the critical value, $2b$, the magnets swing away, one in each direction. There are two equivalent solutions, according to which magnet swings clockwise. The experiment is easily tried, and if you do set it up you will find that before the magnets settle down to their equilibrium configuration in any field they may undergo quite vigorous oscillations, characteristic of two coupled non-linear vibrators. This is irrelevant to the present discussion which concerns itself only with systems in equilibrium, but it is an amusing piece of physics in its own right.

It is worth noting that when $B < 2b$ the arrangement with both magnets exactly parallel to B is a possible equilibrium state, but it is unstable. The continued existence, albeit in unstable equilibrium, of the original configuration marks out the stable symmetric instability from the previous examples of limit point instability, where the original equilibrium configuration disappears entirely.

Exercise 5.10. Consider a magnetically anisotropic crystal suspended, as in Ex. 5.4, on a torsion wire and arranged so that when the wire is not twisted the axis of weaker paramagnetic susceptibility, χ_2, lies along the applied field. Starting from the expressions for the magnetic and elastic energies when the crystal is twisted through θ, show that there is a critical field strength $B_c = [c/m\mu_0(\chi_1 - \chi_2)]^{\frac{1}{2}}$ at which the crystal becomes unstable and may turn in either sense with equal likelihood.

Exercise 5.11. When a coin is spun on a smooth table, it remains upright, spinning about a diameter, until the angular velocity falls below a certain critical value, when it topples and continues

---

* They do not have to be equal, but it saves a lot of mathematical trouble if they are.

spinning (or is it rolling?) with its plane making an angle θ to the horizontal. With further loss of kinetic energy θ decreases until ultimately, and with a characteristic clatter, the coin comes to rest. Here we are not interested in the last stages but in the toppling from a vertical axis. We shall treat the coin as a thin circular disc, rolling without slipping. If it is spinning with its axis at rest before it topples, the centre of the disc remains on the same vertical line, to a good approximation (this is more readily observed by using a larger disc, say 10 cm in diameter, with its rim chamfered to a fairly sharp edge). In fig. 5.9 the centre C remains on the line CN while the point of contact P precesses at angular velocity Ω round the circle of radius $a \cos \theta$ centred on N. From now on we suppose there to be no dissipation, so that C remains fixed in space. Instantaneously the point on the rim that touches the ground at P is also at rest, and the instantaneous motion must be rotation about the line CP, a radius of the disc. Let the angular velocity of this motion be ω. Then the relation between ω and θ can be found as follows:

1. To find how fast the instantaneous axis of rotation precesses, imagine a line of length l drawn normal to the disc through C. The end of the line will trace out a circle of radius $l \sin \theta$ with the precessional angular velocity Ω, and the point itself will move at a speed of $\Omega l \sin \theta$. But we know from the instantaneous rotation about CP that the speed of the point is ωl. Hence $\Omega = \omega \operatorname{cosec} \theta$.

Fig. 5.9. Notation to discuss stability of a spinning disc, in Ex. 5.11.

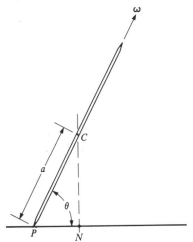

2. Hence show that the horizontal component of angular momentum, $\frac{1}{4}ma^2\omega\cos\theta$, changes at a rate $\frac{1}{4}ma^2\omega^2\cot\theta$.

3. This change of angular momentum is caused by the couple acting on the disc, from which it follows that $\omega^2 = (4g/a)\sin\theta$.

4. The total energy E of the disc, relative to the state of rest, lying on the table, is $\frac{3}{2}mga\sin\theta$. If $E > \frac{3}{2}mga$, no solution of this kind is possible, but the disc can spin with a diameter vertical. At the moment of toppling, when a transition occurs to the type of solution just analysed, $\omega_c = 2(g/a)^{\frac{1}{2}}$.

In this example, as in the previous two, the transition as the speed is reduced does not eliminate the original equilibrium state, but simply makes it unstable. One may observe with a largish disc how the chance vibrations about the steady state get slower and increase in amplitude as ω falls to ω_c. The effect is more easily seen in a closely related instability, that of a disc rolling in a straight line. As it slows down the sideways wobbles become slower and (to conserve their energy) larger in amplitude until they cause the disc to slip and topple over. You need a good area of smooth floor to see this clearly. In this case it is easier to see than with the spinning disc that there are two equivalent patterns of behaviour after the critical condition is passed.

Exercise 5.12. To analyse the stability of a disc rolling along a straight path imagine that its point of contact with the floor wanders slightly about the mean line of the path, with transverse displacement $b\cos kz$, z being measured along the path. At the same time let its centre also wander as $c\cos kz$. It spins with angular velocity ω and has a forward speed of $a\omega$. If there is no slipping the disc rolls parallel to itself, so that the trajectory of its point of contact determines its orientation, while the difference between b and c determines its tilt. It is therefore possible to write down the components of the angular momentum (remember that the principal moments of inertia are $\frac{1}{2}ma^2$ for rotation about an axis normal to the disc and $\frac{1}{4}ma^2$ for rotation about a diameter), and equate their rate of change to the couple acting. It is important not to forget the transverse frictional force at the floor, which provides the acceleration needed for the wavy motion of the centroid. Show that

the frictional force is $-mc\alpha^2\cos\alpha t$, where $\alpha = k\omega r$;

the z-component of angular momentum is $-\frac{1}{4}(b+c)ma\alpha\sin\alpha t$;

hence $(g + \frac{5}{4}\alpha^2 a)c = (g - \frac{1}{4}\alpha^2 a)b$;

the vertical component of angular momentum is

$\frac{1}{4}(ma/\omega)(2\omega^2 c - 2\omega^2 b + \alpha^2 b) \cos \alpha t$;

hence $2\omega^2 c = (2\omega^2 - \alpha^2)b$.

Consequently $\alpha = 0$ or $[\frac{4}{5}(3\omega^2 - g/a)]^{\frac{1}{2}}$. The solution $\alpha = 0$ implies that the disc can roll in a circular arc, while the other solution gives the frequency of lateral oscillations about a straight path. The form of this expression is reminiscent of (2.6), describing the vibrations of an Euler strut. The critical value of ω^2 is $\frac{1}{3}g/a$, so that the slowest speed at which the disc can roll in a straight line is $(\frac{1}{3}ga)^{\frac{1}{2}}$.[4]   17

Having switched from static to dynamical bifurcations in these examples, let us move one step further to the parametric excitation of a passive oscillatory system by changing the value of one of its components periodically at twice the resonant frequency. A specific process of this kind occurs when a pendulum is caused to swing by changing its length periodically.

Exercise 5.13. Fig. 5.10 shows a simple pendulum, which we suppose constrained to swing in the plane of the paper; the wire holding the bob passes through a hole and can be pulled up and

Fig. 5.10. A pendulum excited parametrically by turning the wheel at the top at an angular frequency twice that of the pendulum.

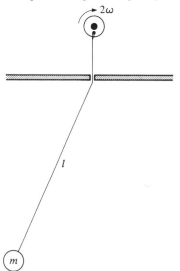

down to excite the pendulum. Show that the tension in the wire varies, as the pendulum swings with small amplitude θ_0, between $mg(1 - \frac{1}{2}\theta_0^2)$ at the ends of the swing, and $mg(1 + \theta_0^2)$ at the middle; so that there is an oscillatory component of tension $\frac{3}{4}mg\theta_0^2 \cos 2\omega t$, if ω is the pendulum frequency.

Now let the wire be pulled up and down by an amount $a \sin 2\omega t$. Show that energy is fed into the pendulum at a mean rate of $\frac{3}{4}mg\omega a\theta_0^2$, and that if τ_e is the time constant for viscous dissipation, the energy gain and loss balance when $a/l = 2/3\omega\tau_e = 2/3Q$.[5]

With $a > 2l/3Q$ the pendulum continues to gain energy exponentially, just like a ball rolling off the top of a curve. This is an example of an unstable symmetric transition. In reality, the damping force on a pendulum bob is rarely proportional to its speed v, but is more nearly proportional to v^2 on account of the turbulent wake it leaves behind. Let us suppose the frictional force to vary as v^n. Show that the mean dissipation rate varies with amplitude as θ_0^{n+1}, so that when $n > 1$ the pendulum is always excited, however small a may be, and that $\theta_0 \propto a^{1/(n-1)}$.

Only if n were 3 (which is not found in practice) should we have θ_0 varying as $a^{\frac{1}{2}}$, the characteristic of a stable symmetric transition. The reality of n being close to 2 results in θ_0 being proportional to a. This is surprising at first sight, since positive and negative a are equivalent, and we expect θ_0 to be a symmetrical function of a. It should be noted, however, that a frictional force proportional to v^2 should really be written as $-v|v|$ so that it is in the opposite sense to v. The appearance of $|v|$ is sufficient to explain any unexpected behaviour at $v = 0$.

Leaving aside the exact development of the oscillation under parametric excitation, we may note that until it is disturbed, by noise or any other interference, the pendulum may stay at rest indefinitely, though in unstable equilibrium. When it does begin to vibrate there is a choice between two equivalent patterns, mirror images, distinguished by the displacement and direction of travel of the pendulum at any particular moment.

Very similar behaviour is to be seen in fig. 4.24, where the bifurcation involves alternating dynamical paths in successive cycles, and where the two equivalent forms of behaviour are also simply antiphased versions of the same thing. One may also regard the transition from focal stability to focal instability, as T changes sign in fig. 2.9, as a bifurcation. In this case, 115 ... 23

however, the new state is not confined to two alternatives, since the phase of oscillation is a continuous variable. It is quite unimportant to try to define exactly what constitutes a bifurcation; the important thing is to understand each example as it occurs, using one's knowledge of other cases if it helps, but not worrying if it does not.

To conclude this catalogue of stable symmetric transitions, consider the experimental arrangement shown in fig. 5.11.[6] When the oil level is just high enough to submerge a little of the square, it is clear that the equilibrium position is with the side of the square horizontal. For if the square is turned through a small angle clockwise, the right-hand side, being more submerged, suffers an upthrust that tends to restore the square to its original position. By contrast, if it is set with a corner pointing straight down, a small clockwise displacement leads to a clockwise couple from the upthrust, and this arrangement is clearly unstable. The argument may be summed up as in fig. 5.11(b), where curve (1) shows how the couple from the upthrust varies with the orientation of the square. Wherever G passes through zero with a negative slope is a stable orientation. It is quite easy to see that at half submergence the pattern is reversed, with the cornerwise orientation (45°) stable, as indicated schematically in curve (2).

As the oil level is increased, how does the symmetry of the pattern change from the 0° orientation to the 45°? If we leave out of consideration,

Fig. 5.11. (a) A plastic square (about 3″ along the sides and $\frac{1}{2}$″ thick is suitable) is freely pivoted on a needle resting on two horizontal ledges attached to the sides of a tank, so that a fraction f would be submerged with a side horizontal. The oil level can be changed to observe how the stable attitude depends on oil level. Thin oil is better than water, as it wets the surface readily. (b) Various possible forms of $G(\theta)$. (c) The theoretical stable positions, showing the stable symmetric transitions from $\theta = 0$ to an asymmetric attitude and thence to $\theta = 45°$. The curves for $0.5 < f < 1$ are mirrors of those shown.

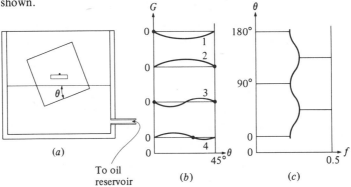

as altogether improbable, that at some intermediate oil level the curve for G might be flat, giving neutral equilibrium in any orientation, the two simplest intermediate forms are those shown as (3) and (4). Let us see what these imply, starting with (3). If this represents the behaviour, the 0° orientation remains stable as the oil rises, until the curve has become flat at 0°, by which time the only stable orientation is at 45°; at this point the square will capsize suddenly and turn all the way to 45°. On lowering the oil level again, it will stay at 45° until the curve is flat there, and then will turn right back to 0°. These transitions are examples of unstable symmetric transitions, with hysteresis, in a bistable system.

The other possibility, represented by curve (4), sees 0° remaining the point of stability until the curve there becomes flat, whereupon the stable orientation migrates steadily across to 45°, reaching there as the curve at that point becomes flat. In this case, which is the one that actually occurs, there is a critical point of bifurcation at which the square begins to turn in either sense, and a second point of bifurcation when 45° is reached, and two possible orientations merge once more. The whole pattern is shown in fig. 5.11(c), and the characteristic form of the bifurcations is clearly displayed. If you feel like working through the theory for yourself, it is not difficult, but it is a tedious business finding the centroid of the submerged portion of the square in order to determine the sign of the upthrust couple.

Rayleigh's principle

Before proceeding to the next category of instabilities, let us note the application of Rayleigh's principle[7] to calculating the critical condition for stability. Rayleigh demonstrated that if one attempts to guess, even roughly, the pattern for the lowest vibrational mode of a system in order to calculate its frequency by the energy method, the answer will always be too high, but not by very much if anything like a reasonable guess is made. Now to determine when a system will become unstable, one must find when the frequency of the lowest mode falls to zero. If we use Rayleigh's method, we shall slightly overestimate the frequency and hence still expect the system to be stable, though the difference should be quite small. In fact we need not go through the whole Rayleigh process, since the condition for a system to vibrate at zero frequency is simply that its potential energy is unchanged by a small displacement – strictly, the quadratic term in the Taylor expansion of the potential energy vanishes. We therefore assume the pattern of displacement and write the condition that V has no quadratic term.

Exercise 5.14. Suppose the Euler strut of fig. 2.5 bends as an arc of a circle, radius R (this is a very bad approximation since the true curvature varies from a maximum at the ground to zero at the load). Show that the strain energy is $EIl/2R^2$ and the potential energy of the load, relative to the undisplaced position, is $-\frac{1}{6}mgl^3/R^2$. Hence the critical load, m_c, is $3EI/gl^2$, which is 21.6% higher than the true value $\pi^2 EI/4gl^2$.

If you are prepared to do a somewhat longer calculation, assume for the shape the simplest polynomial that has zero curvature at the free end, $y = a(3lx^2 - x^3)$. To first order in a^2 the lowering of the load is $\frac{12}{5}a^2l^5$, while the elastic energy, $\frac{1}{2}EI \int (y'')^2 \, dx$, is $6EIa^2l^3$. This gives the critical load as $2.5EI/gl^2$, only 1.3% too high. An assumed sinusoidal shape will, of course, give the correct value for the critical load.

Unstable symmetric transitions

Exercise 5.15. When a long, thin-walled, tube is subjected to external pressure there is a critical pressure at which it is suddenly squashed flat. This is not normally because the material is excessively strained at the moment of collapse (although when a conical glass flask, as used in chemistry, is evacuated and implodes, I think this is due to excessive bending, and hence fracture, at the rim of the base). The instability of the tube is a consequence of the decrease in cross-sectional area when it is deformed into an oval shape, so that the applied pressure does work on the tube. If this exceeds the elastic strain energy the deformation proceeds further and, as experiment shows, becomes catastrophic.

Rather than calculating the correct shape for small deformations, which is a rather long process, let us apply Rayleigh's principle; assume that the cross-section is originally a circle of radius a, and is deformed into the oval $r = a'(1 + \varepsilon \cos 2\theta)$. With thin-walled tubes the critical pressure is insufficient to change the perimeter significantly, and the first stage of the calculation involves determining a' so that the perimeter is $2\pi a$. Show that, to second order in ε, $a' = a(1 - \varepsilon^2)$, so that

$$r = a(1 - \varepsilon^2)(1 + \varepsilon \cos 2\theta),$$

and the area, again to second order, is $\pi a^2(1 - \frac{3}{2}\varepsilon^2)$.* The work

---

* In calculating the area, do not assume the oval to be an ellipse, for which the area is $\pi a^2(1 - 2\varepsilon^2)$. Since the energy changes are all of second order in ε it is necessary to pay much more attention to approximations than in those calculations for which first-order terms are adequate.

done by the applied pressure P, per unit length of tube, is $\frac{3}{2}a^2\varepsilon^2 P$.

Next, calculate the stored elastic energy per unit length, treating a small section of the perimeter as a beam whose curvature is changed. The couple needed to change the curvature C $(=1/R)$ of an already curved beam is $EI\,\delta C$. Hence show that the elastic energy per unit length is $\frac{3}{8}\pi\varepsilon^2 Eb^3/a$, where b is the wall-thickness of the tube.

From these two results, show that the critical pressure is $\frac{1}{4}E(b/a)^3$.

Strictly, the elastic modulus E in this expression is not Young's modulus, as it is in the usual theory of bending beams when Poisson's ratio causes lateral contraction to accompany longitudinal extension. Here, however, there can be no change of dimension along the axis of the tube, though radial changes are not inhibited by the geometry of the arrangement. The appropriate modulus is $E/(1-\sigma^2)$, if E is Young's modulus and σ Poisson's ratio. The latter being typically 0.3 for a metal, the tube is perhaps 10% stronger than the uncorrected formula suggests. The result derived here is well confirmed by experiments on uniform tubes, free from dents which can greatly reduce the critical pressure.

Once collapse has begun, the process can continue with a smaller applied pressure. This is characteristic of an unstable transition. One consequence is that a tube with a kink at one point may begin to collapse there well before the critical pressure is reached, and the collapse will then run along the tube until it is flattened everywhere. If this happens to a length of many kilometres of undersea oil pipeline, about 1 m in diameter, the financial penalty is extremely high.[8]

Exercise 5.16. Rayleigh's principle is also useful in discussing the break-up of a charged conducting droplet, such as a raindrop. Whatever the charge, q, on the drop, the spherical shape is in equilibrium, but there is a critical value of q at which it becomes unstable with respect to elongation into an ovoid. The surface tension T pulls back the pointed end with a greater force per unit area, $T(1/R_1 + 1/R_2)$, where R_1 and R_2 are the principal radii of curvature, smaller at the pointed end; but the extra concentration of charge there gives rise to a compensating outward tension, $\sigma^2/2\varepsilon_0$, if σ is the surface charge density. When this is more than the force of surface tension, the drop explodes.[9] We shall calculate the critical condition by considering energy rather than forces.

Assume the deformed shape of the drop to be a prolate spheroid, with semi-axes a, b and b, where $ab^2 = r_0^3$ to ensure that the volume is the same as that of the spherical drop of radius r_0. The surface energy is T times the surface area S, and

$$S = 2\pi ab[\sin^{-1}e/e + (1 - e^2)^{\frac{1}{2}}],$$

where e is the eccentricity defined by $1 - e^2 = b^2/a^2$.[10] The electrostatic energy is $q^2/2C$, where C is the self-capacitance of the spheroid, $8\pi\varepsilon_0 ae/\ln[(1 + e)/(1 - e)]$.[11] Expand both contributions to the energy as power series in e^2, as far as e^4 (the lowest term in e to appear); show they have opposite signs (positive for the surface energy, negative for the electrostatic energy), and that the critical charge at which they cancel is $8\pi(\varepsilon_0 r_0^3 T)^{\frac{1}{2}}$, when the electrostatic potential of the charge is $2(r_0 T/\varepsilon_0)^{\frac{1}{2}}$. A raindrop of diameter 4 mm would break up if charged to 8000 V above the surrounding potential.

A closely-related instability is that of a drop placed in an electric field. This represents a typical state of affairs in a thundercloud. The drop is elongated in the direction of the field lines as its conductivity allows charges to build up on the surface, and there is a critical field strength beyond which no equilibrium shape exists. This marks a limit-point instability, not an unstable symmetric transition.

In both cases the events that follow the initial instability are similar and remarkable enough to justify a brief mention. The experiments which revealed the behaviour were first carried out by Zeleny,[12] who held a drop on the end of a pipette connected to a voltage source. At the critical voltage the drop deformed into a conical tip from which the liquid emerged in a fine jet. Zeleny's photographs are very striking, and show incidentally that the angle of the cone is close to the value of 98.6° which Taylor[13] showed, much later, to be the only angle at which a charged cone could be in equilibrium, and then only when the electric field had exactly the right strength. No theory of the jet has been worked out.

Exercise 5.17. The spring shown in fig. 5.12 is helical when sufficiently stretched, but on allowing it to shorten it suddenly collapses. The helical form is still possible, but unstable. On the other hand, when the collapsed form is stretched again, there comes a point at which this type of solution ceases to exist, and it jumps back to the helical form. We have here two different transitions, one an unstable symmetric transition and the other a limit point instability. There is marked hysteresis.

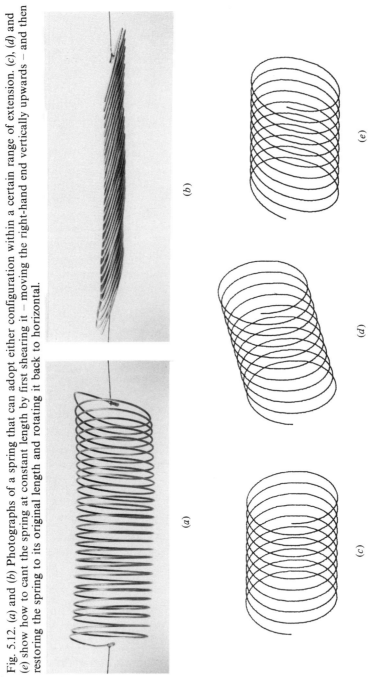

Fig. 5.12. (a) and (b) Photographs of a spring that can adopt either configuration within a certain range of extension. (c), (d) and (e) show how to cant the spring by first shearing it – moving the right-hand end vertically upwards – and then restoring the spring to its original length and rotating it back to horizontal.

(a)

(b)

(c)

(d)

(e)

We discuss here only the condition for the helical form to be stable, as indicated by an increase in strain energy when a slight shear is induced at constant length, so that no work is done in stretching the spring. The calculation is managed conveniently by first shearing the spring as indicated in (c) and (d). The shear involves a second-order extension, and it must be restored to its original length. After rotating its axis back to horizontal we arrive at the deformed shape of (e), whose energy we wish to find. In a fairly tightly wound spring, no great error results from assuming, before it is sheared, that the turns lie normal to the axis. We do not, however, assume that the spring is initially unstrained; we write P_0 for the unstrained pitch, P for the pitch in the state whose stability we examine.

1. If each turn, initially of radius a, is distorted so that the radius varies around the turn according to $a + \delta a \cos \theta$, show that successive turns are deflected sideways by $\pi \, \delta a$. The length of the spring is increased from P to $P + \pi^2(\delta a)^2/2P$ per turn, to first order in $(\delta a)^2$.

2. The strain energy is increased by $\frac{1}{2}\pi EI(\delta a)^2/a^3$ per turn, or $\pi^2 E(\delta a)^2 b^4/8a^3$ for a wire of radius b; E is Young's modulus for the material.

3. The energy removed in restoring the spring to its former length is $\pi^2(\delta a)^2 T/2P$ per turn, if T is the tension in the spring.

4. According to the standard theory of the helical spring,[14] the tension in a spring is $nb^4(P - P_0)/4a^3$, where n is the shear modulus of the material.

5. Combine these results to show that the extra energy needed to shear one turn of the spring, at constant length, is $(\pi^2 b^4/8a^3)$ $[E - n(P - P_0)/P](\delta a)^2$ and the condition for stability is that $(P - P_0)/P < E/n$, which is about 2.6 for steel.

With a normal helical spring, the pitch is increased by applying tension, and $(P - P_0)/P < 1$. The helix is in no danger of instability. The spring in fig. 5.12, however, was originally quite loosely wound, and then was turned inside out (this is quite easily done). In this way it was arranged that it required a considerable pull to start it stretching; $P_0 < 0$ and $(P - P_0)/P \gg 1$. Only by extending the spring was $(P - P_0)/P$ brought down to the value at which the helical form became stable.

Exercise 5.18. A fine jet of water issuing from an orifice breaks up into droplets, as can be seen in fig. 5.13. This is because a slight swelling or necking, if extended over a long enough section of the jet, diminishes the surface area and hence the free energy contribution of the surface.

If the perturbations to the radius are small enough, they can be Fourier analysed into a continuous spectrum of sinusoidal perturbations, each of which behaves independently of the rest. It is therefore convenient to examine a typical Fourier component to see whether it is liable to disappear or to grow; in the latter case the jet will be unstable.

Let the radius vary with length as $a_1 + a_2 \cos kz$. Show that the surface area of one wavelength of the jet is

$$S = (4\pi^2 a_1/k)(1 + \tfrac{1}{4}k^2 a_2^2 - \tfrac{3}{64}k^4 a_2^4 + \cdots),$$

while the volume is

$$V = (2\pi^2/k)(a_1^2 + \tfrac{1}{2}a_2^2), \quad \text{and is constant.}$$

Write V as $2\pi^2 a_0^2/k$, where a_0 is the undistorted radius, and hence eliminate a_1 to show that, up to terms in a_2^4,

$$S = (4\pi^2 a_0/k)[1 - (a_2/2a_0)^2(1 - k^2 a_0^2)$$
$$- 2(a_2/4a_0)^4(1 + 8k^2 a_0^2 + 6k^4 a_0^2)]$$

The quadratic term is negative, indicating instability, if $ka_0 < 1$; and the inherently negative sign of the quartic term shows that we have a case of an unstable symmetric transition.

Exercise 5.19. Let us take this example a little further, by working out the way in which the velocity of waves on the jet depends on their wavenumber, k. It is convenient to use the device of finding how fast the water in the jet must flow in order that waves of a given k shall be stationary with respect to a fixed observer.

Let the water flow at mean velocity v in a jet whose radius varies as $a_1 + a_2 \cos kz$, and assume that the velocity is constant over each cross-section (this is not quite true – see later). Then because of the Bernoulli effect the pressure is greatest where the

Fig. 5.13. Jet of water emerging, on the left, from a long tube of 3.6 mm internal diameter. The jet was vertical, but the photograph has been turned round to save space. The thicker lines are 2 cm apart, and the vertical lines at the top represent the expected separation of the incipient drops. The photograph was taken with an ordinary flash, which accounts for the blurring. Non-linearity is obvious on the right, and the analysis in Exercise 5.19 is valid only for the initial stages.

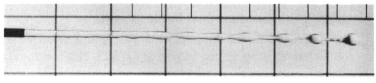

velocity is least. The difference in pressure between air and water is maintained by the surface tension, T. Show that this is possible only if $v^2 = T(k^2 a_1^2 - 1)/2\rho a_1$, where ρ is the density. This gives us the velocity of the wave relative to the jet.

If $ka_1 > 1$, waves can propagate and the mean shape of the jet is not changed; but when $ka_1 < 1$, v is imaginary, so that the frequency is imaginary. Seen by an observer moving with the jet,

$$\omega = vk = k(k^2 a_1^2 - 1)^{\frac{1}{2}}(T/2\rho a_1)^{\frac{1}{2}} = ik(1 - k^2 a_1^2)^{\frac{1}{2}}(T/2\rho a_1)^{\frac{1}{2}}.$$

Then the expression for wave-motion, $\exp i(kz \pm \omega t)$, becomes $\exp(\pm|\omega|t)e^{ikz}$ when ω is imaginary, representing two possible stationary periodic disturbances of radius, growing or diminishing in amplitude with time constant $1/|\omega|$. Normally there will be some amplitude of the growing pattern present, and the diminishing pattern will disappear without preventing instability due to the other. Show that the value of k that yields the fastest growth rate is $1/a_1\sqrt{2}$, corresponding to a wavelength 4.44 times the diameter. In the photograph of fig. 5.13 it is clear that considerable fluctuation in wavelength occurs, as is to be expected since the growth rate is stationary at the maximum, and enough time does not elapse for a very narrow band of wavelengths to dominate. In this example, the lines are drawn above the jet at the theoretical optimum wavelength of $4.44\,d$ and indicate quite reasonable agreement. If the jet is perturbed by an intense sound wave at the appropriate frequency to encourage the optimum wave to form, very regular break-up occurs.[15] This is the idea underlying some ink-jet printers.

Rayleigh,[16] who first gave this analysis, did not make the simplifying assumption that the velocity was constant over each cross-section, but used the correct velocity distribution. The wavelength for fastest growth is changed to 4.51 times the diameter, hardly different from the approximation.

In this analysis we have neglected viscosity. You have only to watch treacle dripping off a spoon to appreciate that it can alter the result. In this case the oscillations are heavily damped, and the unstable modes of low k grow only very slowly.

Influence of imperfections on critical behaviour; the cusp catastrophe

In many of the examples discussed above we have assumed implicitly, or explicitly arranged, that the system shall be ideal. Thus the imploding tube of Ex. 5.15 was assumed to be perfectly round, with no blemish that might favour one mode of collapse over others; and in designing the apparatus shown in fig. 2.4 to study the Euler instability we provided a levelling screw so that the strut had no preference for either direction of bending. In this way we ensured as nearly as possible that the behaviour was modelled by that of a particle in a symmetrical potential:

$$V = -ax^2 + cx^4,$$

in which c was a positive or negative constant. We investigated the behaviour when a was varied steadily from positive to negative, the critical point occurring when $a = 0$.

To introduce asymmetry we may consider adding terms of odd order in x, bearing in mind that in the first instance we are concerned only with the behaviour close to the point of instability; that is, a is to be small and the points where V is minimal are close to the origin, with $|x|$ small. For this reason we do not consider terms of higher order than x^4, which will modify the curve for V only when $|x|$ is large, and concentrate on linear and cubic terms, starting with a linear term εx such as describes the effect of a small tilt applied to the base of the Euler strut:

$$V = \varepsilon x - ax^2 + cx^4; \quad c > 0. \tag{5.1}$$

It is likely, in any application, that the coefficient of x^4 will be found to vary with a, for example, $c = c_0 + c_1 a + \cdots$. Since, however, we are concerned with the primary change that occurs as a changes sign, small changes of c will be of secondary importance, unless by coincidence c and a vanish together (but this will require special treatment). In other words, we shall concern ourselves here, and in what follows, only with the leading terms in the coefficients. Systems which are described by (5.1) have been studied for many years (probably for centuries), and very recently their behaviour has come to be called a *cusp catastrophe*.[17] It will be appreciated that the word catastrophe is used here as a technical term, and need imply nothing like the devastation that its use conjures up in ordinary conversation.

> *Exercise 5.20.* The equilibrium positions of a particle governed by (5.1) are at the points x_0 where $dV/dx = \varepsilon - 2ax + 4cx^3 = 0$. If $|\varepsilon|$ is small enough and $a > 0$ there are three real solutions for x_0, but only one if $|\varepsilon|$ exceeds a certain critical value, ε_c. Show that $\varepsilon_c = (\frac{2}{3}a)^{\frac{3}{2}}/c^{\frac{1}{2}}$. If you do not know the criterion for a cubic equation to have three real roots, it is a good exercise in algebra to work it out for yourself (but always shift the origin of x to reduce it to the standard form without a quadratic term).

This result is illustrated in fig. 5.14, which shows how a large enough value of ε, for a given (positive) a, eliminates one of the minima, and how this critical value $\varepsilon_c \propto a^{\frac{3}{2}}$. Alternatively, if ε represents an intrinsic linear asymmetry, the value of a needed to create two distinct minima is proportional to $\varepsilon^{\frac{2}{3}}$. The characteristically cusped curve of fig. 5.14(c) marks the division between the regimes of one and two minima.

The behaviour can be presented in different ways, and I shall show some of the most useful. First let us fix ε and plot how the positions of the minima and maxima of V vary with a, i.e. the solutions $x_0(a)$ of the equation $dV/dx = 0$ in Ex. 5.20. Rather than solving the cubic equation for x_0, we write $a = (\varepsilon + 4cx_0^3)/2x_0$ and compute $a(x_0)$ as shown in fig. 5.15(a). This may be compared with fig. 4.24, where the imperfection that eliminates the symmetry is a difference between the values of the restitution coefficient at the two walls. The increase with ε of the minimum value of a needed for two stable positions is obvious in fig. 5.15(a), and reflects the result already presented in fig. 5.14(c).

A second presentation shows the value of x_0 as a function of ε and a. The curves in fig. 5.15(b) are drawn for a series of values of a and show $x_0(\varepsilon)$. The meaning can be visualized by thinking of the potential in fig. 5.14(a) as a strip on which a ball is rolling while ε represents a steady force on the ball. With a strong pull to the right ($\varepsilon < 0$), the ball can rest only in the region of positive x_0, at the top left-hand corner of one of the curves in fig. 5.15(b). As ε is made more positive, the ball rolls towards the centre, and x_0 gets less, until the critical moment when its minimum disappears, and it rolls into the region of negative x_0. This is a limit point instability which can be reversed, but only with hysteresis, by decreasing ε again. A typical cycle involving two limit point instabilities is indicated by $ABCDA$. The

Fig. 5.14. To illustrate the effects of imperfections (measured by ε) on a stable symmetric transition. The numbers in (c) indicate the regions where there are two positions, or only one, of stable equilibrium.

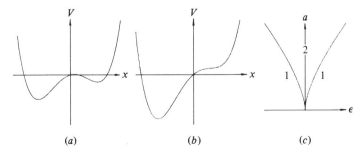

(a)　　　　　　(b)　　　　　　(c)

region of the surface lying between the two instabilities, shown dotted, is unrealizable in practice, since all the extrema of V in this area are maxima, points of unstable equilibrium. If the shadow is projected on to the ε–a plane, as in the diagram, it defines the cusped domain in which two stable solutions exist, the same as fig. 5.14(c).

Bistable systems

The cycle $ABCDA$, involving two limit point instabilities, is an example of the behaviour of a bistable system, having two equilibrium configurations between which it can be switched irreversibly by a suitable perturbation (in this case ε). Bistable systems are common enough, but they are not by any means all examples of a cusp catastrophe; much more frequently they are simply two limit point instabilities. A potential curve like that in fig. 5.16(a) can certainly not be described by a quartic curve (in

Fig. 5.15. Two representations of the cusp catastrophe, as described in the text. Values of ε are indicated on each curve.

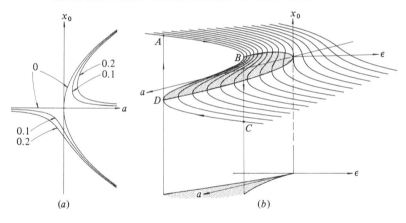

(a)   (b)

Fig. 5.16. Potential $V(x)$ for a bistable system.

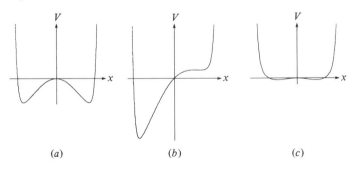

(a)   (b)   (c)

fact it is $-2x^2+x^4+x^{20}$), and might be thought to represent the potential in a toggle switch or a tin lid that can be made to emit loud cracks by pressing on the centre. Indeed, (b) shows how a steady force, from an extra potential $1.5x$, can eliminate one stable position – a ball thus displaced from the right would fall into the left well with a real crump. But one has only to reduce the coefficient of x^2 to $-\frac{1}{2}$, as in (c), to bring the double-well system close enough to the centre for the term x^{20} to be virtually negligible, so far as the limit point behaviour is concerned. Thus the normal bistable system may be taken as a case of a cusp catastrophe far from its critical state. There are, as might be expected, dynamical examples of bistability as well, though here there is no potential function, and the analysis must proceed differently.

Consider, for example, the schematic diagram, fig. 5.17(a) of a positive feedback system (not necessarily an electronic circuit), with amplification A and an inertial element that prevents the signal level at P following that at Q exactly. If this were an electronic circuit with resistance R in the feedback loop, and a capacitance C between P and ground,

$$\dot{V}_P=(V_Q-V_P)/\tau, \quad \text{where } \tau=RC.$$

We take the same equation to represent the behaviour of the feedback loop in other systems, so that V is simply a measure of the signal defining the response.* We also have that $V_Q=A(V_P)\cdot V_P$, and must explicitly take into account that the amplifier can saturate, so that A is less at large input signals. The equation of motion takes the form

$$\dot{V}_P=(A-1)V_P/\tau.$$

Fig. 5.17. (a) A dynamic bistable system, (b) biassed by inserting a cell.

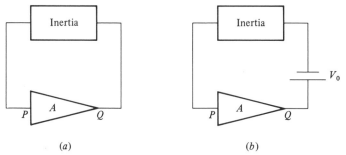

(a) *(b)*

---

* There is a superficial similarity between fig. 5.17 and fig. 2.3 which (minus L) leads to the multivibrator behaviour of fig. 5.8. Note, however, that the saturating amplifier in the latter case controls the resistance of circuit, rather than the loop gain, as here.

If $A > 1$ there is a state of unstable equilibrium $V_p = 0$, but any small disturbance causes V_p to increase in magnitude, either to a positive or negative limit according to the sign of the disturbance; the limiting value of $|V_p|$ is that at which saturation has caused A to fall to unity. By inserting a voltage source V_0 in the feedback loop, as in (*b*), the circuit may be made asymmetrical, and a large enough V_0 will switch it from one state to the opposite.

> *Exercise 5.21.* The circuit in fig. 5.17(*b*) is in equilibrium when $V_P = V_Q - V_0$. If the amplifier behaves like an operational amplifier with negative feedback, it will saturate sharply, e.g. V_Q is equal to AV_P (*A* constant) if V_P is positive and less than V_s, and to AV_s if $V_P > V_s$; and the amplifier may be taken as symmetrical. Draw the relationship between V_P and V_0, to show the range of V_0 in which there are three equilibrium states, and show that the central one is unstable. Discuss also the switching behaviour.

We are obviously far from the smoothness of a cusp catastrophe here. The flip-flop circuit of fig. 5.18(*a*) and the bistable fluidic logic element of (*b*) are realizations of this general scheme. The former needs no comment, but the mode of action of the latter is not so well known, and indeed is

Fig. 5.18. (*a*) An example of a bistable circuit (above), with simplified structure (below).[18] (*b*) A bistable fluidic logic element.[19]

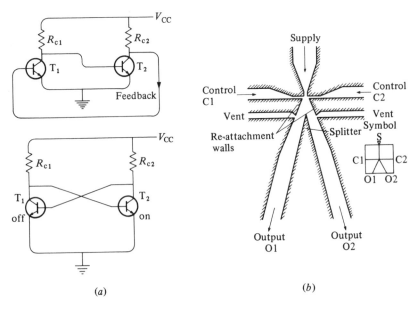

(*a*)   (*b*)

rather obscurely described in most of the few texts that pay attention to the phenomenon.

A jet in an expanding pipe slows down as it proceeds, and the pressure rises, by Bernoulli's theorem.[20] However, near the walls where the motion is controlled by viscosity, the pressure gradient sets up a reverse flow, so that the jet is separated from the walls by a circulation cell. The larger this cell, the more easily fluid can circulate in it, and the more readily it continues to grow so long as the pressure gradient persists. Ultimately the cell on one side wins in the competition, pushing the jet into a region of constant cross-section attached to one side – no matter which – when the pressure gradient disappears and the side of the jet next to the wall is not subject to the danger of a new cell appearing. By injecting fluid from the appropriate control port the jet can be forced off its wall far enough that it reattaches itself on the opposite wall.

The cusp catastrophe (continued)

Returning to the properties of (5.1), let us derive yet another presentation, this time showing how the values of V at the equilibrium positions, $V_0(x_0)$, depend on ε and a. If we were to solve the cubic equation, $(\partial V/\partial x)_\varepsilon = 0$ for x_0, and substitute back in V, we should learn little about the form of V except that it will be triple-valued in the range of ε that permits three real solutions. To compute the cusped curve of fig. 5.19 it is

Fig. 5.19. To illustrate the construction of the equilibrium values of potential $V_0(\varepsilon)$, and the 'Maxwell convention'.

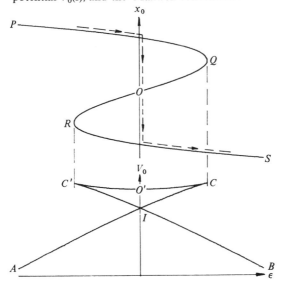

easy to choose x_0 and evaluate $\varepsilon = 2ax_0 - 4cx_0^3$ (see Ex. 5.20), and then to evaluate $V(x_0)$ directly to give $V_0(\varepsilon)$; but to understand why it has this shape a simple trick is helpful. Consider $\int x_0 \, d\varepsilon$, taken at constant a; integration by parts reduces this to $\varepsilon x_0 - \int \varepsilon \, dx_0$, which is immediately integrable to yield the result apart from an insignificant constant:

$$V_0 \equiv V(x_0) = \int x_0 \, d\varepsilon. \tag{5.2}$$

Fig. 5.19 shows how the integration of the many-valued upper curve yields the cusped lower curve. The stretch PQ yields AC in the lower curve, and OQ yields $O'C$. Since $\partial V_0 / \partial \varepsilon = x_0$, the two branches must have the same gradient at C, which is therefore a cusp of zero angle. A similar argument accounts for the rest of the curve. The cusps are present only if $a > 0$; when $a < 0$, V_0 is a single-valued function peaked at $\varepsilon = 0$, but with no crossover.

The range QR describes points of unstable equilibrium, and hence CC' is not stably realizable, but only AC and BC'. The limit point instabilities are at C and C'; on climbing AC (ε increasing) the only thing to do at C is to fall back to the branch BC'. If it is possible to switch from branch to branch there is a continuous path available from A to B, with branch switching at I. This is indicated by the arrowed path on the top curve, and clearly involves switching the system from one minimum of V to the other at the point ($\varepsilon = 0$) when they are equal. The two ways of going from A to B, the irreversible path by way of the limit point C, or the reversible path involving a switch at I, are referred to in some texts as *conventions*, a curious choice of word, suggesting it might be no more than a matter of taste which was adopted. The fact is, one must examine the physical system to discover which course it will take. If it is a ball rolling on a curve it will proceed to the limit point and then roll over to the other side. But a fluid obeying van der Waals' equation has the possibility of proceeding reversibly, by way of a mixture of phases, from liquid to vapour, a process that is the same as the reversible path we have just described.

van der Waals' equation[21]

We introduce van der Waals' equation here as an example of a cusp catastrophe, not as a contribution to the theory of phase transitions which are the main topic of the next chapter. Historically the equation played a central rôle in elucidating the transition between liquid and vapour; almost inevitably it came to be accepted as a valid model of critical behaviour, which it is not. It still has its uses, but must be treated with great caution, as the following discussion should make clear. The details of what really happens at the critical point of a phase transition are very much more complicated than anything described by the theory of the cusp catastrophe. Superficially they may look alike, but to stop at the point reached by van der Waals is to accept a most unphysical description of one of the really important problems in physics still awaiting a full solution. Let us then take van der Waals' equation as a mathematical model showing interesting features in its own right.

Except that they are turned through $90°$, the isotherms of van der Waals' equation, $(P + A/v^2)(v - b) = RT$, in the vicinity of the critical point are very similar to the curves of fig. 5.15(b). (Do not get confused between the van der Waals constant, which I write unconventionally as A, and the coefficient of x^2 in (5.1).) The critical temperature corresponds to $a = 0$, higher temperatures to negative a and lower temperatures, where v is a triple-valued function of P, to positive a. Close to the critical point the isotherms may be approximated by cubic curves (see fig. 5.20(b)). The analogy between van der Waals' equation and the cusp catastrophe of (5.1) is established (if we ignore for the moment a vertical shift as a varies), by equating $P/P_c - 1$ with ε and $v/v_c - 1$ with x_0, P_c and v_c being the critical pressure and volume – $P_c = A/27b^2$, $v_c = 3b$; also $RT_c = 8A/27b$. We write T/T_c as θ, and put $1 - \theta = a$.

Before discussing the analogy further let us look critically at the isothermal transition between liquid and vapour, at any temperature below T_c, which proceeds inhomogeneously (i.e. via a mixture of the two phases) along a horizontal line of constant pressure, such as LV in fig. 5.20(a). According to Maxwell,[22] the areas between this line and the isotherm, one area above and one below, should be equal. This prescription is the same as is involved in the branch switch at I in fig. 5.19. To see this, one must appreciate that in a thermodynamic system in equilibrium the role of V_0 is played by the Gibbs function per unit mass, $g = u - Ts + Pv$.[23] A phase transition can take place reversibly if the value of g for each phase is the same. The horizontal line must therefore be drawn so that $g_L = g_V$. Now $(\partial g/\partial P)_T = v$, so that $g = \int v \, dP$, the integral being evaluated along the isotherm. As in fig. 5.19, this integral generates a cusped curve for g, and the intersection gives the value of P ($\equiv \varepsilon$) that must be chosen to make the integral from L to V vanish. This is Maxwell's condition of equal areas. The cusped curve for g is shown in fig. 5.20(c) for the case $\theta = 0.96$, or $a = 0.04$. Instead of symmetry, like V_0 in fig. 5.19, g rises steadily with P on all branches, as it must since $\partial g/\partial P$ is v and essentially positive.

This argument, however, suffers from a fundamental difficulty which is absent from a simple dynamical system, or one to which (5.1) can be applied validly. In the latter case V exists at all x and one can imagine the system traversing the whole curve in fig. 5.19; thus a ball rolling on a strip can be held at a point of unstable equilibrium by a gentle lateral constraint

Fig. 5.20. (*a*) Isotherms of a van der Waals fluid, showing the critical point at C. The temperatures, $\theta = T/T_c$, are shown beside some of the isotherms. (*b*) Expanded presentation of the isotherms around C, in terms of $x_0 = (v - v_c)/v_c$, $\varepsilon = (P - P_c)/P_c$ and $a = (\theta - \theta_c)/\theta_c$ (shown beside isotherms). (*c*) $g(P)$, being the analogue of the lower curve in fig. 5.19. (*d*) All that may decently be presented of an isotherm and coexistence line.

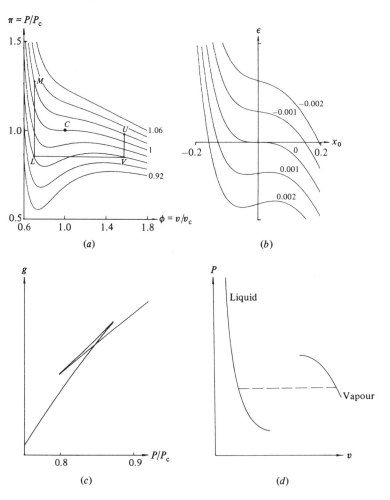

that does not alter V. By contrast, the thermodynamic properties of a gas can be defined only in states of stable equilibrium, and the region of van der Waals' equation between the maximum and minimum of each isotherm is totally unrealizable. The derivation of the equation involves the assumption that, left to themselves, the molecules will spread evenly over the available space (at any rate, on the average). This assumption is fairly valid in the vapour phase, well away from the critical point, and in the condensed phase. It is entirely without foundation in the unstable region between the extrema, as Maxwell of course recognized. There $(\partial P/\partial v)_T$ is positive, so that if any small region chances to contract a little, the pressure therein falls and the molecules outside, exerting a greater pressure, will enforce continued contraction. This is the meaning of unstable equilibrium in an extended system – there are many different perturbations from uniform density possible, and almost all of them will grow uncontrollably, once initiated. We found the same with the liquid jet of Ex. 5.19.

We cannot allow ourselves the escape route that is possible with a simple dynamical system – it would mean holding each molecule more or less fixed while we traversed the unstable region. Leaving aside the practical impossibility, it is actually impermissible since the entropy, and hence g, depend on the extent to which the molecules are free to move. The Maxwell argument is therefore disallowed, unless other considerations can be brought in.

One might hope to get round the problem by making a wide circuit of the critical point and calculating the change of g along it. The path $LMUV$ in fig. 5.20(a) is made up of two constant volume temperature changes, LM and UV, and an isotherm MU, along each of which Δg can be evaluated. This provides a not-too-easy problem in thermodynamics, but you will find if you solve it that to arrive at the Maxwell result it is necessary to assume the thermal capacity C_v to be a function of temperature only. This is indeed a thermodynamical necessity for a substance obeying van der Waals' equation, but only if the isotherm is everywhere realizable. This second attempt fails in the same way as Maxwell's, though the point of failure is better hidden.[24] With any approximate equation of state we do well to draw a typical isotherm as in fig. 5.20(d), containing only such portions as are realizable in principle, and allowing that there must be a coexistence line even if we do not know exactly where to draw it.

The transition from L to V can take place inhomogeneously, once started, by the growth of the vapour phase at the expense of the liquid; this is simply the boiling process. It tells us nothing of the shape of the

175

isotherms between L and V, unless surface tension prevents the formation of a nucleus of vapour in the liquid, in which case superheating can occur, and a portion of the isotherm below L can be realized. Similarly supercooling of the vapour, as in a Wilson cloud chamber, allows a portion of the isotherm above V to be realized.* No amount of supercooling or superheating can succeed in bringing the substance into the unstable range between the extrema, since these are limit points at which an immediate collapse into the other phase must occur.

The cusp catastrophe (for the last time)

We have treated van der Waals' equation as an example of a cusp catastrophe governed by (5.1), but this is not right in detail. For example, the intersections I in fig. 5.19 always occur where $\varepsilon = x_0 = 0$, while in fig. 5.20 the midpoint between the extrema (the nearest we can get to I) occurs at lower P as T is lowered. Also the isotherms are obviously not pure cubic curves. Indeed, (5.1) is an unnecessarily specialized form in which the potential curve is more symmetrical than it need be to generate the characteristic behaviour. The most general form that does not involve higher powers of x than the fourth contains a cubic term as well.

> *Exercise 5.22.* Instead of (5.1), let $V = \varepsilon x - ax^2 + bx^3 + cx^4$, the last two coefficients being taken as constant and describing an asymmetrical well subject to the same modifications by ax^2 and εx as before. Show that this can be reduced to the standard form (5.1) by shifting the origin of x so that $x' = x + b/4c$. Then $V = \varepsilon'x' - a'x'^2 + c'x'^4$, where $c' = c$, $a' = a + 3b^2/8c$ and $\varepsilon' = \varepsilon + ab/2c + b^3/8c^2$. A term in V independent of x' has been dropped as of no consequence.

The effects of the cubic term are, first, that a acquires a constant increment which does nothing to affect the qualitative behaviour, and secondly, that ε acquires a constant $b^3/8c^2$, also unimportant, and a further increment $ab/2c$ that is proportional to a. If now we plot the ε–a

---

* Without the presence of nuclei (dust or charged ions) most vapours can be highly supercooled without condensing, perhaps to a point where the pressure is four or six times the equilibrium vapour pressure. Similarly, many liquids can stand remarkable degrees of supercooling (~ 40 K for pure water in rainclouds[25]) while several simple liquids (CO_2, $SiCl_4$) must be seeded if they are to crystallize at all.[26] It is, however, very difficult to arrange for a solid to be superheated above its melting point. Suitable regions of disorder, or even the free surface, appear to nucleate melting.

curve as in fig. 5.14(c), the general shape will be similar, but at each value of a the midpoint between the branches, instead of being zero, will be $-ab/2c - b^3/8c^2$. The second term merely shifts the curves bodily, but the first shears them, translating the upper parts leftwards if a is positive. Fig. 5.21(a) shows such a case.

This is the sort of behaviour one might expect of an Euler strut which does not start straight; an extreme example is shown in fig. 5.21(c). For any load less than critical a tilt (ε) can be found that maintains the strut fairly well poised, as in the picture. At the critical load itself the correct tilt makes it equally likely to sway either way. The general character of the instability is thus preserved, but, as fig. 5.21(d) shows, the boundary between one and two stable configurations is now sheared; the midpoints of horizontal lines drawn between the branches lie on a line that is not quite straight. This is not surprising, for the complete theory of the strut must contain terms beyond the fourth power, which will disturb the simplicity of the behaviour shown in (a). In view of the extremely bent initial shape it is surprising how small the shear of the ε_c–a curve is.

The van der Waals equation is much more extreme in this respect.

Exercise 5.23. Start with the reduced form of the equation,
$$(\pi + 3/\phi^2)(3\phi - 1) = 8\theta$$
in which $\pi = P/P_c$, $\phi = v/v_c$ and $\theta = T/T_c$, and put $\varepsilon = \pi - 1$, $x_0 = \phi - 1$ and $a = 1 - \theta$. Very close to the critical state, where ε, x_0 and a are all small, show that the equation may be Taylor-expanded to give as leading terms
$$2\varepsilon + 8a + 3\varepsilon x_0^2 + 3x_0^3 = 0.$$
A further expansion gives
$$\varepsilon = -(8a + 3x_0^3)/(2 + 3x_0)$$
$$= -4a + 6ax_0 - 9ax_0^2 - \tfrac{3}{2}x_0^3 + \cdots.$$
Since x_0^2 has a in its coefficient, while x_0^3 has a constant coefficient, it needs only a small shift of the origin of x_0 to eliminate the term in x_0^2 leaving the others unchanged to first order. Hence we write
$$\varepsilon = -4a + 6ax_0 - \tfrac{3}{2}x_0^3 + \cdots.$$
As in Ex. 5.20, show that the condition for the transition from one to three real roots (two stable configurations) is
$$\varepsilon = \varepsilon_c = -4a \pm 8a^{\frac{3}{2}}/\sqrt{3}.$$

This is the curve plotted in fig. 5.21(b). It has the same cusped form as before, but sheared now to run in the general direction of $\varepsilon_c = -4a$. So long

Fig. 5.21. (*a*) The more general form than in fig. 5.14(*c*) of the critical lines separating domains of one and two solutions. The broken line, which is straight, is the locus of the centres of horizontal diameters (but note the remarks in the text about this line in (*d*)). (*b*) The corresponding critical lines for van der Waals' equation. (*c*) A very bent Euler strut, mounted on the arrangement shown in fig. 2.4. (*d*) The critical boundaries were determined by tilting the mount from the poised (two-solution) position until the strut suddenly flopped over. The curves, which are not quite the best fit, are of the general form (*a*). *n* is the number of added strips, as in fig. 2.6; the tilt angle was not calibrated.

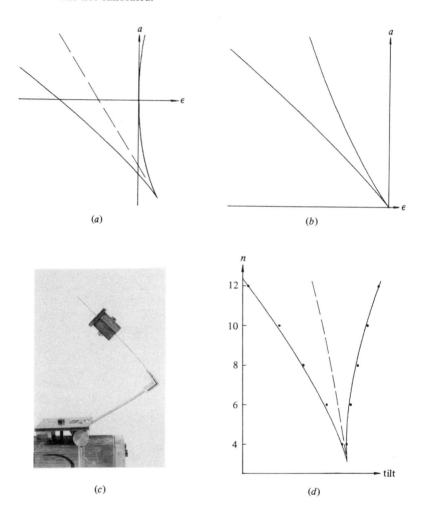

(*a*)

(*b*)

(*c*)

(*d*)

as the potential can be expressed as a cubic polynomial close to a critical point, the cusp must have zero angle.

A system which has several cusps is Zeeman's catastrophe machine, described and analysed by Poston and Woodcock.[27] This provides an example of the topological methods employed by mathematicians engaged in the study of catastrophes.

Other types of catastrophe

The last section was devoted to the cusp catastrophe as it arose when imperfections were introduced into a stable symmetric transition. If the same treatment is applied to the unstable symmetric transition the result is less interesting, as follows from the following argument. The essence of the foregoing discussion is summed up in the form for V,

$$V = \varepsilon x - a x^2 + c x^4,$$

while for the imperfect unstable symmetric transition we should have

$$V' = \varepsilon' x - a' x^2 - c' x^4.$$

Any property of V for a given choice of ε and a is matched in V' by choosing $c' = c, a' = -a, \varepsilon' = -\varepsilon$. Thus when a is negative, there is only one stable configuration and there are no unstable equilibria arising from V; the sign reversal in V' ensures that when a' is positive, there is only one unstable configuration and no stable. Similarly, when a is negative, there is a range of a and ε within which there are two stable and one unstable configurations; hence when a' is positive, there is a corresponding range within which there are two unstable and one stable configurations. There is no possibility of multiple stable solutions and the presence of the imperfection ε leads to no new forms of behaviour.

The same can also be said of the limit point instability (or *fold catastrophe*) whose character is not changed by small perturbations.

[If then we leave the fold and cusp catastrophes, the question arises how much further should we go in discussing higher-order catastrophes? The topological investigations especially associated with the name of René Thom[28] have resulted in a systematic classification which gained great prominence as its protagonists claimed for it a new approach to the understanding of complex phenomena. The actual achievements, however, have been disappointing and sometimes ruthlessly criticized by scientists whose fields of research were the subjects of the clarifying procedures of catastrophe theory. It seems that the conception of what is meant by explanation is a very different thing for a mathematician and for an experimental scientist. It is enough for the former, I judge from Thom's

own book, to have determined into which pigeonhole a given phenomenon is to be assigned. The source of pleasure is more the recognition of how many superficially different observations turn out to be described by the same mathematical structure, than the process of working through the theory in detail to verify that one's knowledge is not just qualitative (important though this may be) but quantitative as well. It is easy to see that the implosion of a tube and the disintegration of a charged water drop (Ex. 5.15 and 5.16) are unstable symmetric transitions; but it demands   150
more than this to apply the results to the design of pipelines or to the understanding of thunderstorms. And this is to leave out of account altogether the genuinely difficult task of following through the processes from the initial stages, as classified by catastrophe theory, to the end as represented by a flattened tube or a Zeleny jet issuing from a Taylor cone.   151

This is said with no intention of demeaning the undoubted success of catastrophe theory as a mathematical invention, but rather as a warning to scientists to be wary of expecting too much from it. This is especially necessary since the mathematics itself is extremely difficult to grasp, and no one should be misled into believing that the great effort needed will immediately put into his hands a powerful tool. In fact most physicists and engineers will go through their lives never meeting in any significant way an example of any catastrophe beyond the two, the fold and the cusp, that I have discussed here. For that reason, I advise no one to attempt to understand catastrophe theory until he has to, or unless he enjoys rigorous intellectual exercise for its own sake; and I shall go into the matter no further.

On the other hand, the examples in this chapter should have made clear that the fold and the cusp pervade the physical world, and that there is good reason to understand their properties fully. But it will be rarely indeed that you will meet a case where the elementary algebra I have employed is inadequate to sort out the essential features. When that happens (e.g. Nye's[29] discussion of caustics in optical systems, or Zeeman's catastrophe machine) you may then find it profitable to understand the topological theory; but a glance at these two examples will show you that the problem must be important to make this worth while, unless you have a genuine mathematical gift. And even then there is a real danger of forgetting the physical problem in your enthusiasm for the logical process.]

6

Phase transitions

The liquid–vapour critical point

The comparison of Figs. 5.19 and 5.20(c), made in the last 161 chapter, left no doubt of the similarity in principle between the gas–liquid 164 phase transition and the cusp catastrophe. At a temperature less than the critical temperature T_c, the curves for g, the Gibbs function per unit mass, intersect at an angle (fig. 6.1) and at this point, where $g_L = g_V$, the two phases may coexist, provided the pressure is right. Since $(\partial g/\partial T)_P = -S$, the entropy is different in the two phases, indicating a latent heat of transition. The high-temperature phase must have the steeper downward slope at the intersection, since it is the phase of lesser g that is stable. Hence the entropy of the transition is always positive – latent heat must be provided to cause the transition from the low-temperature to the high-temperature phase.

All this is as true for a van der Waals fluid as for a real fluid whose isotherms are necessarily discontinuous, as in fig. 5.20(d). The weakness of van der Waals' equation becomes clear when we compare its prediction for the shape of the phase-separation curve with the experimental shape, and the predicted critical isotherm with a real example.

> *Exercise 6.1.* Starting with the expansion, $\varepsilon = -4a + 6ax_0 - \frac{3}{2}x_0^3$, as given in Ex. 5.23, show that the horizontal line $\varepsilon = -4a$ defines 167 the Maxwell construction, and that the line cuts the isotherm at $x_0^2 = 0$ or $4a$. It is conventional to express the relative density difference between the two phases, $(\rho_L - \rho_V)/\rho_c$, as a function of the temperature below T_c. Close to T_c,
>
> $$\frac{\rho_L - \rho_V}{\rho_c} = A\left[\frac{T_c - T}{T_c}\right]^\beta. \tag{6.1}$$
>
> Show that, with van der Waals' equation, $\beta = \frac{1}{2}$ and $A = 4$.

This is plotted in fig. 6.2(a) together with the measured curve for CO_2, whose critical temperature, 304 K, makes it convenient for detailed study.[1] The equation for this curve, corresponding to (6.1), has $A = 3.6$ and $\beta = 0.342$. The lower exponent causes the curve to drop considerably more abruptly as T approaches T_c. Thus at 1 K below T_c the values of $(\rho_L - \rho_V)/\rho_c$ are in the ratio 2.2, but at 1 mK below T_c the difference between van der Waals and real CO_2 amounts to a factor of 6.6. Even so close to T_c the liquid and vapour densities differ by 5%, and careful temperature control is obviously necessary to determine the shape of this curve with any precision.

The experimental problem is further illustrated by the shape of the

Fig. 6.1. The Gibbs function $g(T)$ for two phases at a first-order phase-transition. The stable phases are indicated by heavy lines.

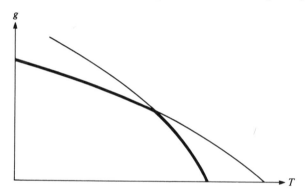

Fig. 6.2. (a) Showing how the density difference between liquid and vapour drops to zero at the critical temperature. The lower curve is the prediction of van der Waals' equation and the upper is the measured behaviour. (b) The critical isotherm on a P–ρ diagram (1) according to van der Waals, (2) as measured for CO_2.

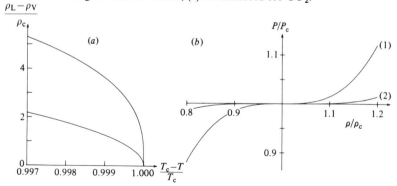

critical isotherm itself (fig. 6.2(b)). From Ex. 6.1 it follows that when $a = 0$, $\varepsilon = -\frac{3}{2}x_0^3$, which means that, very close to the critical point,

$$\frac{P - P_c}{P_c} = 1.5\left[\frac{\rho - \rho_c}{\rho_c}\right]^3$$

according to van der Waals' equation. The measured behaviour of CO_2 replaces this by

$$\frac{P - P_c}{P_c} = \begin{cases} -1.3\left[\dfrac{\rho_c - \rho}{\rho_c}\right]^{4.2} & \text{when } P < P_c, \\[2ex] 1.3\left[\dfrac{\rho - \rho_c}{\rho_c}\right]^{4.2} & \text{when } P > P_c. \end{cases}$$

The critical isotherm is much flatter than van der Waals' equation predicts. It is indeed so flat that its shape may be seriously distorted by even the small pressure variation from top to bottom of a sample chamber, on account of the hydrostatic pressure of the fluid itself. Fig. 6.3 shows how the density should vary with height in a column of CO_2 held exactly at T_c, and with the external pressure just below P_c so that P_c is reached in the middle of the column. A change of height from 1 cm below to 1 cm above this point alters the density by over 10%, while at the centre the density changes by 1% over a height of 0.4 μm. Very shallow sample chambers are needed to get good results, or use may be made of the variation of refractive index with density to probe the critical region optically.[2] It must be remembered, as well, that the extreme sensitivity of $\rho_L - \rho_V$ to temperature demands uniformity and control of temperature to better than 1 mK.

Fig. 6.3. Variation of density (abscissae) with height (ordinates) for a real substance at its critical temperature.

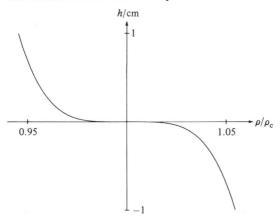

It is not surprising that few substances have been subjected to the same careful measurement as CO_2, but another that has been investigated equally thoroughly is Xe, and its behaviour, plotted in dimensionless co-ordinates as here, is almost indistinguishable. The Law of Corresponding States,[3] which works fairly well over a wide range of physical variables, is particularly well obeyed around the critical point. If we can assume intermolecular attractions to be the same for all molecules, dimensional arguments will show that the law must hold so long as the molecule is not so light (as are He and H_2) that quantum effects enter. The attractions are not exactly the same, however, and this suffices to account for small departures from correspondence. The recognition that these departures disappear as the critical point is approached, and observation of similar behaviour in other types of transition (λ-points), is one of the starting-points for modern theories of phase change. Individual details of intermolecular forces serve to determine the critical constants of a given substance, but the critical behaviour itself is controlled by interactions of larger elements than single molecules, and individual molecular charac-teristics play a much smaller part. *Scaling theories*, using the concept of the *renormalization group*, are extremely difficult to follow and I shall not attempt to give even a summary of their content.[4]

To appreciate why van der Waals' equation fails at the critical point it is helpful to be reminded of a basic assumption in his theory, that the distribution of molecules in space is uniform – there are to be no random fluctuations of density on a large scale, or clusterings on the molecular scale. Clearly as the point of condensation is approached it is unreason-able to expect no clustering; indeed, as we have already seen, one need only trespass beyond the boundary of stability into the regime where $(\partial P/\partial v)_T$ is positive to reach a condition where clustering is inevitable. The critical point, in fact, lies exactly on the boundary of physical realization, where $(\partial P/\partial v)_T$ vanishes and no extra energy (to first order) is incurred in altering the local density.

There are two points here that deserve a little deeper examination. First, the assumption of uniformity in van der Waals' theory; it is not explicit in elementary derivations of the equation of state, but becomes obvious when an attempt is made to relate the constants a and b to the molecular force law. It then becomes clear that the equation can only be formally derived as a first-order approximation in which clustering is neglected. It is something of a paradox that the great historical triumph involved in van der Waals' theory of the continuity of the liquid and gaseous states should have depended on the least justifiable element in its derivation. There is another way of seeing that somewhere in the argument uniformity must have been taken for granted, and that is a purely thermodynamical consequence of the equation.

> *Exercise 6.2.* Show that, for any fluid, the thermal capacity at constant volume, C_v, changes with volume in accordance with the equation
>
> $$(\partial C_v/\partial v)_T = T(\partial^2 P/\partial T^2)_v.$$
>
> To do this, express $(\partial C_v/\partial v)_T$ as $T\,\partial^2 S/\partial v\,\partial T$, reverse the order of differentiation and use one of Maxwell's relations. The result follows immediately. Hence show that for a fluid obeying van der Waals' equation C_v is independent of volume at a given temperature.

The significance of this result in the present argument is obvious. According to classical statistical mechanics the mean kinetic energy of a molecule is $\frac{3}{2}k_B T$, independent of its interactions with other molecules or with anything else at the same temperature. Thus the kinetic energy always contributes $\frac{3}{2}k_B$ per molecule to the thermal capacity. The contribution of potential energy is the only source of departures from this rule so long as the volume is kept constant, and therefore no external work is done (quantum effects may be disregarded for a molecule as heavy as CO_2 and a temperature as high as 300 K). If C_v is independent of volume the effect of a given temperature change on the potential energy must be the same whether the gas is highly rarified or densely compressed. However, if there is any chance of clustering (which certainly changes the potential energy) it will be more marked, and more sensitive to temperature changes, in the dense gas; we conclude that C_v can only be independent of v if no clustering occurs. In a real fluid, C_v does not vary greatly with density, except around the critical point where, entirely at odds with van der Waals, it rises sharply, possibly without limit under ideal conditions.[5]

So much for molecular clustering; the second point concerns fluctuations of density on a larger scale. As long ago as 1863 it was reported that CO_2 at its critical point scattered light (*critical opalescence*),[6] and by 1908 this was recognized, and explained quantitatively,[7] as a consequence of the ease with which density fluctuations could occur when the compressibility was as great as it is near the critical point. The typical volume undergoing density fluctuations must not be much less in extent than λ, the wavelength of light, if any significant amount of light is to be scattered. The way to test whether a given fluctuation is likely to occur is to calculate how much work would be needed to bring it about by an outside agency; if this is of the order $k_B T$ or less, it will easily occur spontaneously – indeed, the whole mass of fluid will be filled with density fluctuations on this and smaller scales. The amount of light scattered from any one region need be only quite small to give noticeable opalescence, since there are so many to contribute to the effect.

> *Exercise 6.3.* Let us try to estimate the conditions for observing a significant amount of light scattering. We start with the assumption that the fluid is separated into globules, about $\frac{1}{2}\lambda$ in diameter, whose refractive indices are $\mu \pm \delta\mu$. The choice of $\frac{1}{2}\lambda$ is purely a guess – a compromise between scattering power and fluctuation probability. Then a ray of light (it is rather dangerous to use ray optics for phenomena whose scale is of the order of one wavelength, but worth trying for the purpose of an estimate) will cross about $6/\lambda$ (can you show this?) interfaces between globules in the course of travelling unit distance through the fluid. If a fraction $(\mu_1 - \mu_2)^2/(\mu_1 + \mu_2)^2$ of the energy is scattered at each interface, as in Fresnel's theory of reflection at normal incidence by an interface,[8] the fraction scattered in unit length is given by
>
> $$\alpha \sim 6(\delta\mu)^2/\mu^2\lambda.$$

α is called the *extinction coefficient*.

Next we must estimate the variations in μ to be expected. The first thing to show is that it takes energy $K(\delta v)^2/2v_0$ to compress a small volume v_0 by δv, K being the isothermal bulk modulus $-v(\partial P/\partial v)_T$. So long as this energy is not much more than $k_B T$ it can be provided temporarily from thermal energy. We therefore take $(\delta v/v_0)^2 \sim 2k_B T/Kv_0 \sim 30k_B T/K\lambda^3$, if v_0 is the volume of a sphere of diameter $\frac{1}{2}\lambda$. Now $\mu^2 - 1$ is proportional to density in a not-too-dense fluid, so that $|\delta\mu/\mu| = [(\mu^2 - 1)/2\mu^2]\,\delta v/v_0$. Hence

$$\alpha \sim 45(\mu^2 - 1)^2 k_B T/\mu^4\lambda^4 K.$$

A proper calculation by Ornstein and Zernike[9] gave a very similar form to α, but with rather heavier scattering,

$$\alpha = (8\pi^3/3)(\mu^2 - 1)^2 k_B T/\mu\lambda^4 K. \tag{6.2}$$

The extra factor of about 4 probably means that the size of the fluctuating regions that contribute most to scattering is nearer $\lambda/3$ than $\lambda/2$. There is nothing unlikely about this.

We now use (6.2) to find the conditions necessary to observe strong opalescence, and if we take α to be $1\ \mathrm{m}^{-1}$, so that the scattering in a path of

1 cm should be readily seen, and $\mu = 1.2$ for CO_2 at its critical point, we find K must be about 10^6 N/m^2. Examination of the critical isotherm indicates that it is not a very severe requirement to reduce K to this value – the density must be right within 10% and the temperature within 1 K – and indeed opalescence is observed over a temperature range of several degrees.[10] It is important, however, to give the fluid a long time to reach equilibrium, since the thermal capacity has a high peak at the critical point, even C_v for which van der Waals' equation predicts no peak at all. The high value of C_v makes equilibration slow. For this reason typical demonstrations in which a sealed phial of CO_2 is allowed to cool fairly rapidly through T_c may show strong opalescence, but this is simply due to phase separation. Although the conditions under which Andrews was working when he first observed the effect would have made true critical opalescence possible, it must be admitted that the description he gives raises doubts. But the mistake, if mistake it was, was undeniably fruitful, and in any case his full study of the isotherms of CO_2 remains a classic of experimental physics.

To sum up the discussion so far, the qualitative picture of the critical point behaviour, at the point where the vapour–liquid equilibrium line terminates, is very similar to a cusp catastrophe, of which van der Waals' equation is an example. In the near neighbourhood of the critical point, however, the quantitative description ceases to have the elementary analytical form of a cusp catastrophe, and the critical point itself is non-analytic – the properties cannot be described by a Taylor expansion around it. Since van der Waals excludes density fluctuations, which we have seen become prominent near the critical point, it is a reasonable conjecture that inclusion of the fluctuations would change the detailed behaviour and might provide an explanation for the failure of analyticity. This is easier said than done – indeed, no complete theory has been developed; but there are approaches to the problem that are fairly readily understood, and I shall set out one of these in due course. It is easier, however, to see what is happening on the basis of different sorts of critical point, and the first task is to recognize some of the varieties provided by nature, and to see how far one may go towards explaining them with the help of catastrophe models analogous to van der Waals'.

Lambda-point anomalies in the thermal capacity

Fig. 6.4 shows examples of a rather frequent phenomenon, the appearance of a sharp peak in the otherwise smooth temperature-variation of the thermal capacity. Pride of place must go to liquid ^{4}He which below 2.17 K is a superfluid, and which was the occasion of the name λ *(lambda)-point* for this type of behaviour; in the event it has turned out to be quite different in its origin from the other examples named after it, and we shall discuss it no further. It was by no means the first lambda anomaly to be noted, but the first to be studied seriously, and it is not so long ago that it began to be recognized that the general properties they share are more significant than individual differences in detail.

The overriding point in common is that there is no latent heat at a λ-transition. The entropy changes continuously, and there is no analogue of two-phase coexistence such as we find in first-order transitions. Of course, if the sample being studied is so non-uniform that the λ-point occurs at different temperatures in different parts there will be a narrow temperature range in which some regions are above the λ-point and some below. But in a substantially perfect sample like liquid ^{4}He (if we ignore the very slight effect of gravity) the transition from the normal liquid to the superfluid, below the λ-point, affects the whole sample simultaneously. Under ideal conditions a λ-transition is homogeneous, in contrast to the heterogeneous first-order transition, for which a nucleus of the new phase must form and will thereafter grow at a rate controlled by the heat supply.

Fig. 6.4. λ-transitions: (a) ^{4}He ($T_\lambda = 2.17$ K).[11] Note how changing the temperature scale raises the curve without altering its shape very noticeably. C_s is the thermal capacity of the liquid in equilibrium with the vapour.

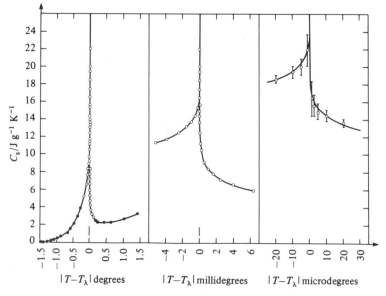

(a)

As long ago as 1944 Onsager[17] had given a complete mathematical analysis of a special model of a λ-transition (the two-dimensional Ising model) which showed the thermal capacity C rising to infinity at the critical temperature T_c, albeit very slowly, as $\ln|T_c - T|$. In 1957, Buckingham *et al.*[11] measured the thermal capacity of liquid helium to

Fig. 6.4. (*b*) Fe at its Curie temperature;[12] (*c*) β-brass, CuZn, at its order–disorder transition;[13] (*d*) NaNO$_3$ at its structural transition.[14]

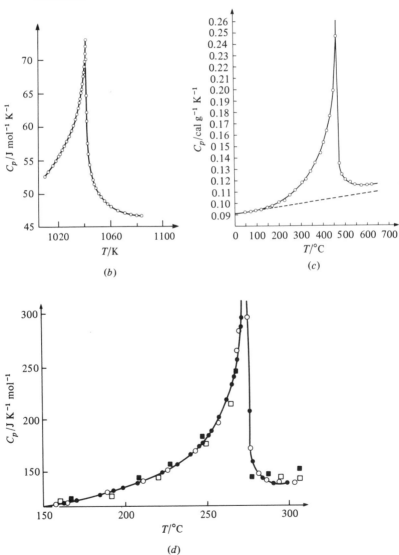

(*b*)

(*c*)

(*d*)

within 10^{-6} K of its λ-point and found close correspondence to Onsager's result. Nothing quite so striking has been achieved with solids because of residual non-uniformity; nevertheless there is no reason to doubt that for many substances perfect samples (and perfect measurements!) would show C increasing without limit. It should be mentioned that in a number

Fig. 6.4. (*e*) $MnCl_2 \cdot 4H_2O$ at its antiferromagnetic Néel point;[15] (*f*) Sn at its superconducting transition.[16]

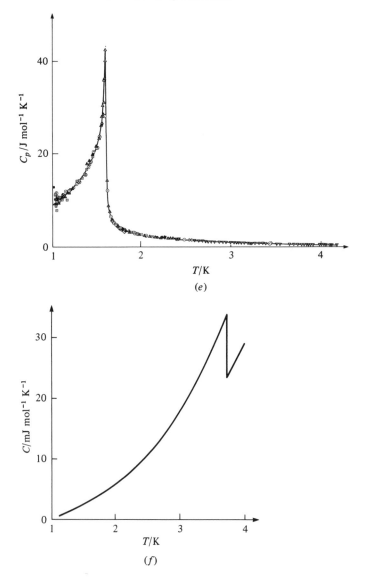

(*e*)

(*f*)

of solids (quartz, NH_4Cl) the transition has been thought, even after careful measurement, to be a λ-point only for this ascription to be rejected when still more careful examination revealed a small latent heat just as C seemed to be set for a continuous rise to infinity. Perhaps this point needs more precise definition – in a true λ-transition the thermal capacity, though rising to ∞ at T_c, is well behaved in the sense that

$$\int_{T_c-\varepsilon}^{T_c+\varepsilon} C\,dT \to 0 \quad \text{as } \varepsilon \to 0;$$

but if there is a δ-function at T_c, so that the integral tends to a non-zero limit L, L is the latent heat and the transition is of first-order. It is a vexed question why, and under what conditions, the λ-point should degenerate into a weak first-order transition,[18] and I do not propose to discuss it, but confine attention to true λ-points. This, however, can probably be asserted safely, that the first-order transition with latent heat is found only in C_p, measured under constant pressure, and would not appear if C_v were measured. It is a very general observation that an anomaly in thermal capacity appears in more extreme form in C_p than in C_v; for example, a van der Waals' fluid has no anomaly at all in C_v at the critical point, while C_p goes to infinity.

A quick glance at fig. 6.4(a) to (e) might suggest that the width of the peak, relative to T_c, varies widely, but this is rather misleading since some peaks, for example (b), (c) and (d), are superimposed on a high base-line contributed by lattice vibrations, while the low temperature peaks (a) and (e) account for almost the entire thermal capacity. It may be noted that a peak of the Onsager form, $\ln |T_c - T|$, has no width – if you expand the temperature scale to see the region close to T_c more clearly, the shape is unchanged and the curve is merely lifted bodily, as illustrated in (a). The area under the curve for ΔC, the additional part of C due to the λ-anomaly (if that can be unequivocally determined), or strictly $\int (\Delta C/T)\,dT$, tells one the entropy change associated with the anomaly. For many λ-transitions this is fairly close to $R \ln 2$, as if something in each molecule which has two degrees of freedom above T_c is frozen into immobility on cooling. The entropy change is a valuable diagnostic tool in seeking the mechanism responsible for the anomaly, but this is hardly our business. Be warned, however, not to take the figure of $R \ln 2$ as more than a rough indication – it is by no means a universal constant and indeed many transitions involve a much smaller entropy change. The ferromagnetic transition, as in (b), is such a case, and there are low-temperature ferromagnets like Ni_3Al whose anomaly is hard to detect; a superconductor (f) shows only a small discontinuity. Nevertheless there are theoretical reasons for believing that a perfect sample of a superconductor would show a λ-anomaly within 10^{-10} K of T_c, but we shall have long to wait before this can be tested.

Second-order transitions

For a time, especially during the 1930s, it was customary to regard a transition like fig. 6.4(f) as typical of λ-anomalies, because there was a reluctance to believe (until Onsager's paper of 1944) that C could rise to infinity. The anomaly was envisaged as a finite discontinuity in C, so that the assumed variation of entropy with temperature was smooth and differentiable, except for a sharp, but finite, discontinuity of slope at T_c. Consequently it could fairly be regarded as an analytic function of, say, T and P, except that the coefficients in the Taylor expansion of $S(T, P)$ would be different above and below the phase separation line $T_c(P)$. Such transitions have become known as second-order transitions. It is now clear from experiment and theory that this is too much of a simplification. The presumed infinity in C at T_c implies that $S(T)$ momentarily runs vertically, as in fig. 6.5(b). As a result, S is continuous but not differentiable at T_c, which is a singularity; that is, S cannot be expanded as a Taylor series about T_c. Thus any theory, such as Landau's,[20] which assumes (two-sided) analyticity is fundamentally flawed. It nevertheless has considerable heuristic value as a guide to the varieties of behaviour that may be expected; moreover, it can be extended to include non-analytic features and so point the way to better treatments. Leaving aside the question of phase transitions, Landau's model provides an excellent illustration of stability theory, and its very failure in detail when applied to phase transitions suggests where to look to appreciate the limitations of analytical catastrophe theory in this case.

Fig. 6.5. The variation of entropy around a transition point. (a) Second-order transition as assumed by Landau. (b) A true λ-point. (c) First-order transition with latent heat.

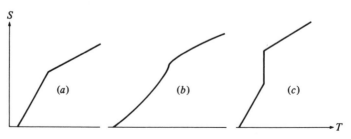

Landau stressed that as a system is cooled through its λ-point it begins to develop some new property that can be quantitatively measured. Here are a few examples:

*Order–disorder transition in β-brass, CuZn.** Well below T_c copper and zinc atoms occupy alternate positions on a body-centred cubic lattice to form two interlocking simple cubic lattices, one occupied by copper atoms and the other by zinc. As the temperature is raised some copper and zinc atoms exchange places, so that a number of Cu-sites are occupied by Zn, and the same number of Zn-sites by Cu; above T_c the occupation is random and the alloy completely disordered. That description, at any rate (or something like it) underlies the Landau viewpoint, and the degree of order can be defined as the difference between the number of copper atoms and the number of zinc atoms on one set of sites, divided by the number of sites:

$$\eta \equiv (n_{Cu} - n_{Zn})/(n_{Cu} + n_{Zn}).$$

Since the two sets of sites are equivalent, an arrangement of all copper atoms on Zn-sites, and vice versa, is indistinguishable from what we described as perfect order, and $\pm \eta$ therefore represent the same degree of order. Accordingly we expect any analytical expression for the thermodynamic properties to depend on η^2 and contain no power of η.

Before proceeding to the other examples of λ-points let us examine the β-brass case more closely. It is clear that copper and zinc prefer to alternate, i.e. that the energy is minimized by surrounding each Cu by Zn and each Zn by Cu; a certain amount of energy must be supplied to effect an exchange when the system is totally ordered, $|\eta| = 1$. On the other hand, in the disordered state each atom's eight nearest neighbours will, on the average, consist of four Cu and four Zn. An exchange now costs no energy, since there is nothing to say which position is right and which wrong. As the temperature is raised, then, from well below T_c, more atoms get displaced and it becomes easier for still more to follow suit. Nevertheless complete randomness can only be expected at a very high temperature, since there is always some lowering of energy, even though it may be much less than $k_B T$, to be obtained by surrounding a copper atom by zinc. Around any chosen copper atom we expect to find the immediate neighbourhood reasonably well ordered, so there is a good chance of

---

* Unfortunately the ideal alloy, with exactly equal numbers of copper and zinc atoms, is not stable and recrystallizes into a mixture of different species, but the stable almost-ideal β-brass behaves so closely like one expects for the ideal composition that we can pretend it is the ideal we study.

finding copper atoms at the nearest positions on the single-cubic lattice of Cu-sites. Moving to the next sites beyond those diminishes our expectation of finding Cu atoms, since if one of the nearest sites has probability p of holding Cu, the next nearest will have a probability which is not, perhaps, exactly p^2 but something of that order. Thus, as we proceed further from our first-chosen Cu, the memory of whether it was Cu or Zn begins to fade until, far enough away, we are as likely to find Cu as Zn. This is what happens above the λ-point; there is no *long-range order* such as would give Cu a preference over Zn however far away we look, but there is *short-range order* defined by a correlation length – the distance one has to go before a substantial measure of randomness has set in (this concept can be precisely defined, but we shall not need more than a rough picture). As we approach the λ-point from above, the correlation length goes to infinity, but it is only as T_c is passed that long-range order sets in.

Now let us start from 0 K and proceed upwards towards T_c. At very low temperatures a few Cu–Zn pairs will be found to have exchanged places in the otherwise ordered arrangement. A simple exchange, however, only slightly disturbs the environment and certainly gives little encouragement to other pairs in the vicinity to make the exchange – the energy required will still be much more than $k_B T$. As T rises, however, each exchanged pair begins to act as a nucleus from which disorder spreads a little way into the ordered environment, and the distance it spreads becomes progressively further the more disorder there is and the less energy is needed for exchange. In this way we form a picture which is a sort of mirror image of the picture above T_c – local disorder, i.e. $|\eta|$ locally smaller than the mean value $(\bar\eta = |\eta(T)|)$, with the difference $\bar\eta - |\eta|$ falling to zero as one moves to greater distances from the nucleus of disorder; the characteristic correlation distance for disorder rises to infinity as T approaches T_c from below.

It should not be supposed that in the neighbourhood of T_c, where the correlation lengths are large, one can simply identify the nuclei of disorder $(T < T_c)$ or order $(T > T_c)$, since the spheres of influence of individual nuclei overlap. One now must envisage a fluctuation of η from point to point, with the correlation length defining the typical distance over which η can change substantially. A complete theory of the thermodynamic properties must describe not only the mean value of η and its temperature dependence but the amplitude of the fluctuations and their correlation length. Just as van der Waals assumed perfect uniformity of density, so here the earliest theories assumed the degree of order to be uniform, and its value, η, to be the only significant parameter. It was soon realized, however, that the tail of the peak in C, persisting above T_c (see fig. 6.4) indicated that there was not complete disorder as soon as the λ-point was passed; entropy, i.e. disorder, still has to be created as the temperature rises, to break up the islands of short-range order, and the thermal capacity reflects this need. Onsager's[17] complete solution of a two-dimensional order–disorder problem, which fully described all the fluctuations and their correlation length, was a masterpiece of mathematical physics which has never in the forty succeeding years been extended to three dimensions. Some, indeed, who have worked for years on the problem have concluded that no mathematical methods exist powerful enough to cope with it.

The other typical examples of λ-transitions can be dealt with briefly:

Ferromagnetism in Fe, Co, Ni, Gd and many alloys. Above the critical temperature (called in this case the *Curie temperature*) the substance is paramagnetic, acquiring a magnetic moment per unit volume, M, parallel and in proportion to the strength of an applied field. The susceptibility, $\lim_{H \to 0} (M/H)$, tends to infinity as the temperature is lowered to T_c, and below T_c the substance possesses a spontaneous magnetic moment even in the absence of an applied field. This statement requires qualification; normally, different regions (*domains*) are spontaneously polarized in different directions, with the result that the overall moment is small. Only a rather weak field is needed, however, to bring all the domains into alignment, the strength of field required being much less than would be needed to change the spontaneous moment significantly. I shall return to the question of domain formation later in the chapter. It is enough at this stage to note that below T_c the spontaneous zero field moment rises sharply according to the law $M \propto (T_c - T)^\beta$, where β is close to, but probably not exactly, $\frac{1}{3}$. The Landau theory predicts $\beta = \frac{1}{2}$, the difference being analogous to that already remarked on, following (6.1), for the liquid–vapour critical point.

The spontaneous magnetization has its origin in the spin magnetic moment of the electrons. Usually the spins are randomly oriented but may be partially ordered by applying a strong magnetic field – this is spin paramagnetism such as is shown above T_c in ferromagnetics, greatly enhanced by the intrinsic tendency of the spins to take up parallel orientations. In the highly idealized Ising model, the electrons occupy lattice sites and have the choice of two opposite spin orientations, the interaction between neighbours being such that they tend to line up parallel. Formally this model is analogous to the order–disorder model in which at each lattice site there is a choice between Zn and Cu. In the light of the analogy we can accept the difference in number between up and down spins, in other words, the spontaneous magnetization M, as an order parameter analogous to η. The Bragg–Williams theory[21] of the order–disorder transition and the Weiss theory[22] of ferromagnetism are both primitive descriptions of phase transitions which turn out to be formally identical; both assume η (or M) to be uniform and both arrive at results which accord with the general Landau theory.

Ferroelectricity.[23] This is the electrical analogue of ferromagnetism, and occurs in Rochelle salt, KH_2PO_4, $BaTiO_4$, etc. In barium titanate the positively charged barium ion sits in a cubic cell formed by negative titanate ions. The cell is rather too big for the Ba ion, which rattles around inside it at high temperatures with a preference for the corners, where the energy is lowest. On cooling towards T_c the preference begins to become marked, and when an ion settles near a corner it tries to distort its cell into a slightly sheared cube, so that the neighbours are encouraged to follow suit. At T_c itself, long-range order appears for the first time – the whole crystal develops a spontaneous dipole moment from the asymmetric disposition of the barium ions, and at the same time the crystal ceases to be cubic as a systematic shear sets in. It is convenient to choose the spontaneous electrical polarization as an order parameter, but the shear angle would do just as well. There are, indeed, many crystals (e.g. $NaNO_3$ – see fig. 6.4(*d*)) which undergo shear transitions, accompanied by λ-anomalies, without acquiring a spontaneous dipole moment, and for these the shear angle is the best available order parameter.

One may see an analogy between these *structural* phase transitions and a stable symmetric transition, e.g. the Euler strut. The spontaneous shearing of the lattice, with either sign of the shear angle, is like the flopping of the strut to the right or left. And just as the restoring force for lateral displacements of the strut goes to zero at the critical load, so the shear modulus associated with the particular transition drops with temperature, more or less linearly, reaching zero close to T_c. The same analogy is found in the ferromagnetic phase transition, where the inverse of the magnetic susceptibility (the stiffness of the individual moments against lining up by an applied field) falls to zero as the temperature is approached at which no external help is required for alignment.

14

132

Landau theory

We may now tackle the Landau theory, which assumes that the partially ordered state can be characterized by a single-order parameter, η, fluctuations being disregarded. If the sample could be kept disordered even below T_c (this is not inconceivable in, say, β-brass where the ordering process is quite slow enough to allow thermal measurements to be finished before much ordering has taken place) the free-energy, F, of a sample kept at constant volume, as we shall assume throughout, will surely be a smooth function of temperature, $F_0(T)$. When ordering takes place, however, F is changed, and Landau assumes that the change is analytic in η. Moreover, in such cases where η and $-\eta$ describe equivalent ordering patterns, the change will be represented as a power series in η^2, rather than η:

$$F(T,\eta) = F_0(T) + \alpha\eta^2 + \tfrac{1}{2}\beta\eta^4 + \cdots, \tag{6.3}$$

in which α and β are smooth functions of temperature. At each value of T, η will take a value that minimizes F, i.e.

$$\eta = 0 \quad \text{or} \quad (-\alpha/\beta)^{\frac{1}{2}}. \tag{6.4}$$

Above T_c, in the disordered state, $\eta = 0$ must be the sole stable solution; and since for stability F must lie at a minimum, it follows that $\beta > 0$. If $\alpha > 0$, no other solution exists, but new possibilities arise when α becomes negative. The transition temperature is thus identified as the point where α passes through zero. When $\alpha < 0$, it is the solutions $\eta = (-\alpha/\beta)^{\frac{1}{2}}$ which are stable. The leading role in the transition is played by α, and so long as we consider only temperatures near T_c, β may be taken as constant, while

$$\alpha \sim \alpha'(T - T_c), \tag{6.5}$$

α' being assumed constant and positive.

Substituting in (6.4) and (6.3), we find

$$\eta = 0 \quad \text{or} \quad [\alpha'(T_c - T)/\beta]^{\frac{1}{2}}, \tag{6.6}$$

and

$$F(T) = \begin{cases} F_0(T); & T > T_c \\ F_0(T) - \alpha'^{\frac{1}{2}}(T_c - T)^2/2\beta; & T < T_c. \end{cases} \tag{6.7}$$

As soon as we plot $F(T)$ the resemblance to fig. 5.3(c) is obvious; the Landau theory is the thermodynamic analogue of the stable symmetric transition. In some cases it is possible to go further and illustrate the cusp catastrophe, but for this purpose one must be able to destroy the symmetry of $\pm\eta$. Thus if η represents M, the magnetization of a ferromagnet, application of an external field lowers the minimum for M parallel to H, and raises that for M antiparallel to H. There is no need to pursue the analogy further – it is quite obvious from the discussion of the cusp catastrophe; and the magnetization curve $M(H)$ follows automatically from the earlier treatment. The fact that for some λ-transitions there is no way of biassing the minima in the way that H does for a ferromagnetic (or E for a ferroelectric) does not, of course, destroy the analogy as far as it goes.

It is already clear that according to this theory $\eta \propto (T_c - T)^{\frac{1}{2}}$, a prediction that can be fairly easily tested with a ferromagnetic, and found to give something like the experimental behaviour though certainly not a quantitatively correct description. Similarly the thermal capacity is easily predicted, and tested experimentally without too much difficulty; here, too, is a discrepancy, perhaps even more obvious. From (6.7) the entropy $S(T)$ follows by a single differentiation, since $S = -(\partial F/\partial T)_v$:

$$S(T) = \begin{cases} S_0(T); & T > T_c, \\ S_0(T) - \alpha'^2(T_c - T)/\beta; & T < T_c. \end{cases}$$

Hence

$$\begin{aligned} C_v = T(\partial S/\partial T)_v = C_0; & \quad T > T_c, \\ C_0 + \alpha'^2 T/\beta; & \quad T < T_c. \end{aligned} \tag{6.8}$$

It is predicted that as T rises through T_c, S will be continuous but with an abrupt change of slope (fig. 6.5(a)) and C_v will drop suddenly by an amount $\alpha'^2 T_c/\beta$, a phenomenon observed only in the superconducting transition (fig. 6.4(f)). There is no hint here of the typical λ-anomaly as in the other examples in fig. 6.4, and indeed they could not be expected to come out of any discussion that assumes analyticity, as in (6.3).

The fact that the details of the critical behaviour are wrongly predicted does not destroy the value of Landau's analysis, whose great strength is the hope it gives that a wide variety of different critical phenomena can be treated as instances of a single unifying concept, the order parameter. If by itself it is insufficient to characterize the thermodynamical functions, but must be supplemented by a specification of the range and magnitude of the fluctuations in η, this does not invalidate belief in a unifying principle, but simply indicates that more than one parameter is needed to specify the material in each case. There are other positive achievements as well, as may be learnt from Landau and Lifshitz's book.[20]

Weiss theory of ferromagnetism[22]

This section is an interpolation outlining a specific microscopic model which led, long before Landau's general theory, to a most fruitful description of the Curie point behaviour of a ferromagnet. Naturally in 1907 Weiss worked with a strictly classical model, but the quantum modification only simplifies the argument without changing its essentials. It is assumed that the material contains unpaired electrons whose magnetic moments, of magnitude μ_B, are free to take either of two orientations, parallel or antiparallel to any magnetic field that may be present. Even in the absence of an applied field, according to Weiss, each electron experiences a field produced by the others, and proportional to the average net magnetization. A dimensionless constant of proportionality λ is introduced to describe the interaction: if the electrons combine to contribute a mean moment M, per unit volume, each electron finds itself in a field $\lambda\mu_0 M$, relative to which it may orient its own moment, μ_B, either parallel or antiparallel, so that it has available two energy levels $\pm\lambda\mu_0 M\mu_B$. These are occupied in accordance with the Boltzmann distribution, a fraction $a/(1+a)$ being parallel and $1/(1+a)$ antiparallel, where $a = \exp(2\lambda\mu_0 M\mu_B/k_B T)$. Hence the moment of unit volume, containing N electrons, is given by

$$M = N\mu_B \tanh(\lambda\mu_0 M\mu_B/k_B T),$$

or

$$zT/T_c = \tanh z, \tag{6.9}$$

where

$$z = \lambda\mu_0 M\mu_B/k_B T \quad \text{and} \quad T_c = \lambda\mu_0 N\mu_B^2/k_B.$$

Exercise 6.4. (a) Solve (6.9) graphically by plotting $\tanh z$ against z, and determining where it is intersected by the straight line $-zT/T_c$ for various values of T/T_c. Note that $M = 0$ is the only solution when $T > T_c$. Satisfy yourself, by considering the effect of a small perturbation, that non-zero values of M are stable, while 0 is unstable, for $T < T_c$. Alternatively, plot the solution to (6.9) by selecting a series of values of z and calculating the value of T/T_c to which each corresponds.

(b) By expanding $\tanh z \sim z - \frac{1}{3}z^3$, show that just below T_c

$$M \sim (3Nk_B/\lambda\mu_0)^{\frac{1}{2}}(T_c - T)^{\frac{1}{2}}, \tag{6.10}$$

having the same dependence on $T_c - T$ as (6.6), derived from Landau's theory.

(c) Show that the contribution to the total energy U, due to the distribution of electrons on the two levels, is $-N\lambda\mu_0 M\mu_B$ $\tanh(\lambda\mu_0 M\mu_B/k_B T)$; hence that the thermal capacity, dU/dT, has a discontinuity of $3Nk_B$ at T_c.

(d) By comparing these results with (6.4)–(6.6), write (6.3) in the form

$$F = F_0 + \lambda\mu_0(T - T_c)M^2 + \lambda^2\mu_0^2 M^4/6Nk_B.$$

(e) The saturation moment of iron at a temperature well below T_c shows that it has 2.22 electrons per atom contributing to the moment, or 1.87×10^{29} m^{-3}, each electron possessing a moment of 1 Bohr magneton, $\mu_B = 9.27 \times 10^{-24}$ JT^{-1}. The Curie temperature of iron is 1143 K. Hence calculate that $\lambda = 1460$.

It should be noted that (d) demonstrates how the coefficients in Landau's expansion of F are related to the microscopic parameters, but in no intuitively obvious way. Thus it is not until one has fully solved the Weiss model to obtain T_c that one can discover the meaning of Landau's α and its proportionality to $T - T_c$. Landau in fact makes a wide generalization to the effect that so long as a single-order parameter characterizes the thermodynamic behaviour, all analytic microscopic theories, whatever their atomic mechanisms, must lead to an equation like (6.3).

A second interesting point from (e) is that although Weiss could be satisfied that the theory explained the occurrence and general features of ferromagnetism quite well, it was only by introducing a parameter λ whose large value was quite inexplicable in the context of his physics. The only source of an internal magnetic field known to him was that produced by the individual moments, and these could not have caused λ to be greater than about unity. Heisenberg's discovery of the exchange phenomenon[24] in quantum mechanics provided a mechanism whereby a neighbouring pair of electrons could interact strongly, as if each was generating a strong magnetic field to act on the other; only 'as if', for the exchange force has its origin in the Coulomb interaction and Pauli's exclusion principle, not in any magnetic process.

Apart from the magnitude of λ the most important difference between Weiss' and Heisenberg's approaches lies in the latter's explicit recognition that it is only near neighbours that influence one another, not the average magnetization as assumed by Weiss. The short-range exchange force allows fluctuations in the local value of M to be included in the theory, if the mathematical problems can be solved. It is to one attack on this problem that we now turn.

Ginsburg–Landau theory of fluctuations[25]

The basic assumption both of van der Waals and Landau theories is that of uniformity – a single parameter suffices to define the internal configuration. Both theories lead to analytical expressions for the thermodynamic functions at the critical point, in conflict with the indications of experiment and the only exact theory, Onsager's treatment of the two-dimensional Ising lattice. The so-called Ginsburg–Landau theory (its historical origins are hard to sort out, but this name is as fair as any) attempts to account for the singular behaviour without a full description of the fluctuations of order, by concentrating on the extended fluctuations to the neglect of those taking place over a short range. There is a good physical reason for this: in principle only an infinite system can show a genuine singularity, and one must look to a size-dependent process if one hopes to get at the truth. In a finite system the number of energy levels of the system as a whole may be extremely large, but it is finite. In principle the thermodynamic properties may be calculated by forming the Partition function,[26]

$$Z = \sum_i e^{-\varepsilon_i/k_B T}, \tag{6.11}$$

the sum being taken over all energy levels of the system, whose volume is to be kept fixed. Then the free energy, F, is derived from Z,

$$F(V, T) = -k_B T \ln Z, \tag{6.12}$$

and from F the other thermodynamic functions follow; for example, $S = -(\partial F/\partial T)_v$, $C_v = -T(\partial^2 F/\partial T^2)_v$. So long as the number of terms in the sum (6.11) is finite, Z is finite and differentiable, and so is $\ln Z$. It follows that C_v cannot be singular, though there is no objection to its being very sharply peaked.

Now let us imagine a very large sample divided into many identical blocks, each containing a considerable number of atoms. Each block by itself would behave rather like a large sample except that its free energy would be smoothed to some extent. Clearly if we imposed on each block the requirement that it should at all times exhibit precisely the same degree of order as all the rest, the sample as a whole would behave as a larger copy of the individual block, with a free energy exactly the same as that of the block, but multiplied by the number of blocks. It is when the blocks have the ability to behave as individuals that the extra degree of freedom introduces extra entropy, and allows the free energy of the whole to assume a sharper form at the λ-point. The Ginsburg–Landau theory admits to possessing little knowledge of the free energy of a single block, and devotes itself to the extra contribution due to the variation of behaviour between blocks. What it does know about the blocks is that their free energy is analytic and should therefore be expressible, close to T_c, in the form (6.3):

$$F_{block} = F_0 + \alpha\eta^2 + \tfrac{1}{2}\beta\eta^4, \tag{6.13}$$

where η is now the mean value of the order parameter in the single block.

The state of the sample is now defined by assigning a value of η to each block, and it is assumed that the total free energy is not merely the sum of terms like (6.13), but has an additional contribution due to the interaction of neighbouring blocks. If two neighbours have values of η, η_1 and η_2, differing from the mean η_0 which minimizes (6.13) the extra free energy is taken as proportional to $(\eta_2 - \eta_1)^2$; a positive constant of proportionality ensures that the most likely configuration is such that all have $\eta = \eta_0$. Small fluctuations, however, enough to increase F by no more than a few $k_B T$, are to be expected. It should be noted that only η enters this expression for the interaction energy – all states of a block corresponding to a given mean degree of order interact identically. This is an assumption, the simplest but not necessarily correct.

It is convenient to treat η as a continuous function of position, remembering that it has a meaning only at the centre of each block. Then the interaction energy has density $\gamma |\mathrm{grad}\ \eta|^2$. It is also helpful to write η as $\eta_0 + \delta$, assuming $\delta \ll \eta_0$, and to express the spatial variations of δ in the form of a Fourier series

$$\delta(r) = \sum_k \delta_k \cos k \cdot r,$$

just as atomic displacements and momenta are expressed in, for example, Debye's theory of specific heats.[27] The total number of Fourier components is equal to the number of blocks in the sample. In a large sample, of volume V, as in Debye's theory, the number of components in the range k to $k + dk$ is $V k^2\ dk/2\pi^2$. Any one component contributes interaction energy $\gamma k^2 \delta_k^2$ per unit volume, and the contributions of each combine to give the total interaction energy, with no correcting terms so long as the quadratic form $\gamma |\mathrm{grad}\ \eta|^2$ is taken as exact. It should be noted that γ is assumed to be sensibly constant around T_c, and certainly not to go to zero with η. This assumption is justified by the study of models of ferromagnetism and superconductivity for which γ can be related to the microscopic properties.

We are now ready to calculate the free energy of the whole sample. First let us ignore the interaction energy and consider the free energy of the assembly of independent blocks. If (6.13) is interpreted as the free energy of a single block, per unit volume, the effect of a single Fourier component is to change the equilibrium value, with $\eta = \eta_0$, to

$$\bar{F}_{\text{block}} = F_0 + \alpha \overline{(\eta_0 + \delta_k \cos k \cdot r)^2} + \tfrac{1}{2}\beta \overline{(\eta_0 + \delta_k \cos k \cdot r)^4}$$
$$= F_0 + \alpha \eta_0^2 + \tfrac{1}{2}\beta \eta_0^4 + \tfrac{1}{2}(\alpha + 3\beta \eta_0^2)\delta_k^2, \tag{6.14}$$

a term in δ_k^4 being neglected. And since, according to (6.6), $\eta_0^2 = \alpha'(T_c - T)/\beta$ when $T < T_c$ (the only case we shall consider), (6.14) may be rewritten

$$\bar{F}_{\text{block}} = F_0 - \tfrac{1}{2}\alpha'^2(T_c - T)^2/\beta + \alpha'(T_c - T)\delta_k^2. \tag{6.15}$$

Now we can add in the interaction energy and sum over all Fourier components to give the free energy per unit volume of sample:

$$F(T, \delta_k) = \Phi(T) + \sum_k \frac{1}{2} \mu_k \delta_k^2, \tag{6.16}$$

where

$$\Phi(T) = F_0 - \tfrac{1}{2}\alpha'^2 (T_c - T)^2/\beta \quad \text{and} \quad \mu_k = 2\alpha'(T_c - T) + \gamma k^2.$$

It is important to remember that (6.16) expresses the free energy when the amplitudes of the Fourier components are constrained to take special values δ_k. When the constraint is removed, the δ_k are free to take any value and will indeed change all the time. To find the free energy of the unconstrained system, let us return to (6.12) and note that the Partition function for a given choice of δ_k takes the form

$$Z(\delta_k) = e^{-\Phi/k_B T} \exp\left[-\sum_k \frac{1}{2} \mu_k \delta_k^2 / k_B T \right]$$

$$= e^{-\Phi/k_B T} \prod_k \exp\left[-\mu_k \delta_k^2 / 2k_B T \right]. \tag{6.17}$$

Before reduction to the form (6.17) Z was the sum of exponential terms each of which had in its exponent the energy of a quantum state of the whole system, subject to the choice of δ_k. The complete Partition function is a sum with the constraint removed, and is obtained by integrating over all the variables δ_k.* It is a fortunate circumstance that the variables separate in (6.17), and that each integral is a readily evaluated Gaussian integral. Without this separation the Ginsburg–Landau approach would come to a halt at this point. As it is, we note that

$$\int_{-\infty}^{\infty} e^{-z^2} \, dz = \pi^{\frac{1}{2}},$$

so that

$$Z = e^{-\Phi/k_B T} \prod_k (2\pi k_B T/\mu_k)^{\frac{1}{2}}.$$

Finally, we use (6.12) again to recover F, taking the opportunity to drop all constants that will not contribute to $\partial^2 F/\partial T^2$, to substitute for μ_k from

---

* The Partition function, as defined by (6.11), can only be uniquely evaluated for a quantized system having definite energy levels, ε_i. To apply the argument to a continuous distribution such as the values available to δ_k, we divide the range of δ_k into many equal elements, and sum over all. When the sum is replaced by an integral, as here, the size of the element appears as a multiplier, or as a term proportional to T in F. On differentiation the entropy is found to have an additive constant, so long as we assume the size of the element to be independent of temperature; and the value of C is unaffected by the choice of element. It is rash to ask why the size of the element should be temperature-independent.

(6.16), and to integrate over k with the weight factor $k^2/2\pi^2$, appropriate to waves in a three-dimensional box. Then

$$F = \Phi_0 - \tfrac{1}{2}Nk_B T \ln T$$
$$+ (k_B T/4\pi^2) \int k^2 \ln\left[2\alpha'(T_c - T) + \gamma k^2\right] dk. \tag{6.18}$$

The second term is the sum of contributions $\tfrac{1}{2}k_B T \ln T$ from each Fourier component, N in number. Its meaning is clarified by noting that it provides a term in the thermal capacity, $-T\,\partial^2 F/\partial T^2$, equal to $\tfrac{1}{2}Nk_B$; this is what Boltzmann's equipartition law demands for N particles, without kinetic energy, confined to quadratic potential wells. We have here the consequence of finding terms proportional to δ_k^2 in (6.15), which causes each component to contribute the amount appropriate to one quadratic degree of freedom. This result makes it possible to estimate the amplitude of each component, since $\tfrac{1}{2}\mu_k\overline{\delta_k^2} = \tfrac{1}{2}k_B T$. Hence

$$\overline{\delta_k^2} = k_B T/[2\alpha'(T_c - T) + \gamma k^2]. \tag{6.19}$$

Now the fluctuations of η in any one block are the resultant of all the Fourier components acting at the centre of the block. Since the components are independent they must be compounded with random phase, and according to a well-known rule their amplitudes add quadratically:

$$\overline{\delta^2} = \sum \delta_k^2 \rightarrow \int k^2\overline{\delta_k^2}\, dk/2\pi^2, \tag{6.20}$$

the integral being taken up to the largest k needed to make up a total of N components, one for each block in unit volume of material. If $T = T_c$, (6.19) takes a simple form enabling (6.20) to be evaluated as $k_B T k_{max}/2\pi^2\gamma$. It is now obvious that the Landau mean field theory can never retain its validity right up to T_c; as T_c is approached η_0 falls to zero, while the fluctuations in η do not. Ultimately it is the fluctuations in order, rather than the degree of order itself, that dominate the thermodynamic behaviour.

As for the behaviour of C as T_c is approached, we must look at the last term in (6.18), noting that differentiation with respect to temperature may be carried out under the integral sign, and also that the upper limit of integration is finite, so that no singular behaviour can arise here. It is at the lower limit, $k \rightarrow 0$ for a sample of unlimited size, that divergence is possible. We therefore concentrate attention on the term which has the highest power of k in the denominator, and which determines the nature of the singularity:

$$C_{sing} = (\alpha'^2 k_B T_c^2/\pi^2) \int k^2 \, dk/[2\alpha'(T_c - T) + \gamma k^2]^2. \qquad (6.21)$$

The integral is clearly divergent when $T = T_c$.

> *Exercise 6.5.* Make the substitution $z = [\gamma/2\alpha'(T_c - T)]^{\frac{1}{2}}k$ to show that as $T \to T_c$ the upper limit of integration may be set at infinity, and that in the limit $C_{sing} \propto (T_c - T)^{-\frac{1}{2}}$.

Although the Ginsburg–Landau theory predicts an infinite peak in C at T_c, the exponent, $-\frac{1}{2}$, is at variance with the measured exponent of -0.12 ± 0.01 for iron. The theory is, in fact, only a first approximation to a complete theory, but it does at least make clear the crucial role of long-wavelength fluctuations; or, to put it another way, it shows how the correlation length diverges at T_c. We shall take the discussion no further. Not only does it become extremely difficult, but we have achieved our immediate purpose, which is to recognize that λ-transitions (and the liquid–vapour critical point) are non-analytic and cannot be described by a finite number of internal parameters. They fall completely outside the range of catastrophe theory. Some who have extolled the generality of catastrophe theory as a classification of the varieties of physical experience appear to have overlooked this fact.

Domains

The theory of λ-transitions does not usually concern itself with imperfect samples, having small variations of purity, stress, etc., in different parts. In reality, not every region of the sample has its local λ-point at exactly the same temperature, and this has several consequences; for one thing, the transition is slightly smeared, and the predicted infinity in C is replaced by a high, but finite peak. Secondly, as the sample is cooled slowly through T_c, different regions become ordered first, probably in ways that do not match. In β-brass, for example, we may expect the first sporadic islands of order to expand and merge, so that the Cu-sites in one region are on the same simple cubic lattice as the Zn-sites in another. In the end, instead of perfect order permeating the crystal, there will be regions of virtually perfect order separated by phase–antiphase boundaries, where the occupation of the sites is reversed. Similarly, in a transition in which a cubic lattice becomes slightly tetragonal, we may expect to find the tetragonal axis lying along any one of the three original cubic axes, with a different choice in different regions. At the boundaries of regions such as this there will be an extra surface energy due to the mismatch of the ordered patterns in neighbours. All this, however, is an accidental consequences of imperfect samples, of little theoretical interest for the pure scientist, however much they may attract the attention of those concerned in the production of perfect crystals, or material scientists for whom imperfections are an essential factor determining technical behaviour.

On the other hand, however perfect a ferromagnetic or ferroelectric crystal may be, it will almost always contain domains in which the permanent moment points in different directions. This is a completely different matter from the accident of imperfection. A large single crystal sample of iron, magnetized uniformly by applying a weak field along a cube axis (which is a preferred direction of magnetization), will not retain its magnetization when the field is removed, but will once more become a mosaic of differently magnetized domains. Otherwise the external magnetic field produced by the sample itself, acting like a permanent magnet, would greatly increase the energy of the whole, by an amount $B^2/2\mu_0$ per unit volume in the space outside. By forming many differently magnetized domains the external field energy can be greatly reduced, though at the expense of surface energy at the domain boundaries. Here we have a new sort of problem in stability, to discover the domain configuration that will minimize the total energy.* The technical importance of iron has assured prolonged and detailed study of domain patterns, such as are well described in specialized texts.[28] I shall do no more than indicate some of the simpler examples to show the principles involved, confining attention to single crystals.

The form of the domains within a sample, unless it is a thin film or plate, must usually be inferred from the arrangement they take up at the surface. This can be beautifully revealed, as in fig. 6.6, by electropolishing the surface (to give flatness and polish without strain) and painting on a thin layer of fine magnetic powder in colloidal suspension; the attraction of powder to the field lines at domain boundaries gives rise to the Bitter patterns, called after the inventor of the technique.[29] The sample in fig. 6.6(*a*) was polished to give a face parallel to a cube face of the iron crystal. Every domain has its magnetization parallel to one or other of the easy directions of magnetization, vertical and horizontal, in this example. The arrows showing the magnetization vector make clear how 'free poles', i.e.

---

* Strictly the free energy, but the distinction is not of great importance to the argument.

Fig. 6.6. Domain arrangements in iron single crystals as revealed by the Bitter technique.[29] For description see text.

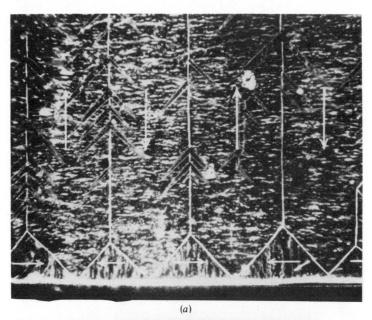

(a)

(b)

divergence of the magnetization, is avoided – neighbouring domains are either antiparallel, or, at the surface, closure domains eliminate surface polarisation and meet the internal domains at 45°. In the latter case the normal component of magnetization is continuous across the boundary.

The little imperfections in the diagram presumably arise from defects in the surface, e.g. pinholes, and are but a foretaste of how the pattern is changed when the surface is no longer parallel to an easy magnetization direction. The 'fir-tree' pattern of fig. 6.6(b) is a typical example of the intricacies involved in the minimum energy configuration. The extra energy associated with any free poles is so large that it pays to have smaller domains, and hence more surface energy at the domain walls, if thereby the amount of free polarity can be reduced. A fuller discussion would involve quantitative estimates of the interface energy and the field energy due to divergence of M, and I shall not go into the problem any further.

A very similar type of problem arises in superconductors, which when cooled below their critical temperature show a first-order phase transition at a certain critical value of applied magnetic field. In the analogous case of liquid–vapour equilibrium below T_c, liquid and vapour may coexist in almost any configuration at the equilibrium vapour pressure. In a superconductor, however, the interface between the normal and the superconducting phases must lie parallel to the local magnetic field, since no field may enter a superconductor. This forces the mixed (*intermediate*) state into something like a laminar pattern, with the laminae lying parallel to B. The problem arises of minimizing the extra energy of the distorted field just outside, without paying too big a price in extra interface energy; closure domains are not possible in this case.[30]

Critical slowing-down

Microscopic processes in condensed matter normally take place very rapidly; after a system has been disturbed from equilibrium by a step-function of, say, electric field or pressure, it relaxes to a new state of equilibrium in a time τ which is frequently much shorter than 1 μs. Thus the imaginary (lossy) part of the compliance, measured at frequencies in the MHz range, is often found to vary as ω^2, an indication that $\omega\tau \ll 1$. It is not unusual to find a substance that otherwise shows little loss becoming very lossy in the vicinity of a λ-point. The explanation is that the relaxation time increases sharply as T_c is approached, probably to infinity in most cases. Not all response functions acquire substantial imaginary parts, but only those which are appropriately coupled to the microscopic variable involved in ordering. Take the λ-transition in ^{4}He, for example.

Whatever the detailed explanation may be, there is no doubt that the equilibrium degree of order is affected by pressure – the variation of λ-temperature with pressure shows this. When a step-function of pressure is applied, therefore, internal changes will be needed to restore the state of equilibrium. If these take longer near T_c than elsewhere, the attenuation of ultrasonic pressure waves will reveal it. In fact, except near T_c the attenuation is very slight, but it rises very much like the thermal capacity, to a high peak at T_c, so that a disturbance at 1 MHz is virtually lost after travelling a few centimetres. On the other hand, the electrical permittivity, which is determined by the atom alone and not by the configuration of its neighbours, remains real, with no hint of dielectric loss, at all temperatures, and all frequencies from optical downwards.

A single example will serve to indicate why processes slow down near T_c. There is no need to consider real λ-transitions – the Landau theory will serve our need. We start by extending (6.3), as applied to a ferromagnet, by introducing an external field B_0 and in consequence an extra term $- B_0 \cdot M$ in the free energy:

$$F = F_0 + \alpha M^2 + \tfrac{1}{2}\beta M^4 - B_0 \cdot M. \tag{6.22}$$

If $T > T_c$, so that $\alpha > 0$, and if B_0 is weak enough, M will remain small, and the term in M^4 may be disregarded. It follows immediately that F is minimized when $M = B_0/2\alpha$. The volume susceptibility, κ_m, defined as $\mu_0 M/B_0$ in the limit $B_0 \to 0$, takes the value $\mu_0/2\alpha$, or $\mu_0/2\alpha'(T - T_c)$. This is the Curie–Weiss Law, which states that κ_m rises to infinity as the temperature falls to T_c.

Now let us look at this from a microscopic point of view, imagining the magnetization to be due to localized electrons whose moments can point only along the line defined by the direction of B_0, either parallel or antiparallel. When $B_0 = 0$, electron moments tumble back and forth as they are acted upon by the thermal motions of the lattice, and they have no preference for either direction. When B_0 is applied, however, the rate of tumbling towards the direction of B_0 is slightly increased, while the opposite rate is slightly decreased, since energy is given out in the one process and taken in in the other. This has already been illustrated for uncoupled moments in chapter 3, where it was shown that after a step-function of B_0 the magnetization relaxes exponentially to its new equilibrium. Curve (*a*) in fig. 6.7 shows this response function. We now introduce the interaction term αM^2 in (6.22), and repeat the step-function process, starting from $B_0 = 0$. Immediately after B_0 is applied $B_0 \cdot M$, being linear in M, dominates αM^2, so that the response function starts at the

67

same gradient as before. As M, however, continues to increase, where previously the energy difference of the up and down states was enough to compensate for their different populations, αM^2 brings the two states more closely together, and allows M to continue rising along curve (*b*); ultimately it will saturate at a higher level, much higher indeed if T is very close to T_c, and κ_m much larger than the susceptibility of uncoupled electron moments. If the response function remains exponential, as the following exercise shows, and the tumbling rate is sensibly temperature-independent, it follows that the relaxation time is proportional to κ_m, i.e. to $(T-T_c)^{-1}$.

> *Exercise 6.6.* Modify the model used in chapter 3 to apply to a
> ferromagnetic obeying the Weiss theory, i.e. instead of a level
> spacing of $2\mu_B B_0$ write $2\mu_B(B_0+\lambda\mu_0 M)$. Show that when $T > T_c$,
> M relaxes exponentially to its equilibrium value, which is
> enhanced by a factor $(1-\lambda\mu_0 N\mu_B^2/k_B T)^{-1}$, and that the
> relaxation time is enhanced by the same factor.

67

188

It is not to be expected that the elementary Landau theory will give a perfect description of the temperature-dependence of τ. Even so, measurements[31] of the attenuation of ultrasonic waves in liquid ^{4}He, just below the λ-transition, indicate that τ is well described by $8.8 \times 10^{-12}/(T_\lambda - T)$ s to within 5×10^{-5} K of T_c. At that temperature τ has risen to the point where $\omega\tau \sim 1$ at 1 MHz, the frequency of the measurements, and the simplifying assumption that τ is proportional to the attenuation constant no longer holds. There is no reason to believe that measurements made at

Fig. 6.7. Illustrating critical slowing down. The step-function responses start with the same slope, but (*b*), which has further to go than (*a*), takes correspondingly longer.

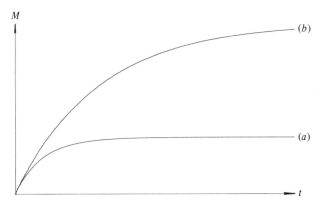

a lower frequency (which would be rather difficult) would not show τ continuing to rise as $T_\lambda - T \to 0$.

The dielectric loss can also be used to study τ for such materials whose dielectric constant rises to infinity as the transition to a spontaneously polarized ferroelectric state is approached. Measurements[32] on $NaNO_2$ showed the electrical susceptibility χ varying as $(T - T_c)^{-1.11}$, and $\tau \propto (T - T_c)^{-1.3}$; in a Landau theory, both exponents would of course be unity. It must be added that $NaNO_2$ is not an ideal material, since as it is cooled it first becomes antiferroelectric and then, a few degrees below, ferroelectric. Nevertheless, one would expect $\tau \propto \chi$, whatever the details of the temperature dependence of each, while the experiment suggested $\tau \propto \chi^{1.18}$. All the same, the general idea of critical slowing-down is well vindicated.

Superficially this is similar to other time effects we have noted at critical points, and we may pause to enquire into the validity of any analogy we may be tempted to see. Take, for instance, the stable symmetric transition 132 represented by a potential $V = -ax^2 + cx^4$; a particle, of mass m, moving in this potential has a frequency for small oscillations such that $\omega^2 = 2|a|/m$ if $a < 0$, $4a/m$ if $a > 0$. As a goes through zero, the period of oscillation goes to infinity as $|a|^{-\frac{1}{2}}$. This critical slowing down, however, refers to oscillations about equilibrium, not relaxation accompanied by entropy production as the free energy tends to a minimum. The analogy is improved by immersing the dynamical model in so viscous a fluid that the inertial term in the equation of motion is negligible. A particle in a parabolic potential $V = \frac{1}{2}\mu x^2$, subject to a retarding force $-\lambda\dot{x}$, relaxes exponentially to the origin with time constant λ/μ. In this case, then, $\tau = \lambda/2|a|$ for $a < 0$, $\lambda/4a$ for $a > 0$; the time constant rises as $1/|a|$ towards the critical point. This indeed is a fair analogy, since the Landau equation is the equation of a stable symmetric transition, and the relaxation process for the individual moments results in exponential approach to equilibrium, as in the viscous model. If you draw the step-function response of the model to an applied force (not omitting now the mass of the particle), you may find the comparison with fig. 6.7 still further strengthens the analogy.

Two other instances of processes becoming slow at a critical point are (1) the transition from focal stability to focal instability on crossing the line 23 $T = 0$ in fig. 2.9 and (2) the impulse oscillator (or any system bifurcating to chaos) at a point of bifurcation. In (1) the time constant for decay of 115 oscillations rises to infinity, while in (2) it is the time required to settle down to a steady state. You may like to develop the analogy between these and a λ-transition, but I do not recommend you to try too hard. An analogy is useful if it enables ideas from one field to produce fertile results in another; if it is harder to understand the analogy than either of the processes being compared, the only value is the intellectual exercise involved.

Nevertheless, whether the mechanisms are to be regarded as similar or not, it remains a general tendency for the time-scale of a process to stretch out as a critical point is approached.

7

Broken symmetry

Symmetry changes at a critical point

It was a central feature of Landau's exposition of his general theory of phase transitions that a change of symmetry occurs at a critical point. He came to give the concept of symmetry a broader meaning than is customary but this need not concern us, since we have already had enough illustrations in which the properties above T_c are isotropic or, more usually, reflect the crystal symmetry; in particular a sample reacts in the same way to an applied field B and to its opposite, $-B$. Below T_c, however, the appearance of a permanent moment gives a ferromagnetic sample a polarity which it does not possess above T_c, and it is differently affected by $+B$ and $-B$. This is obvious if the sample as a whole possesses a magnetic moment, and is turned in opposite directions by $+B$ and $-B$, but it is equally true when the domain structure results in zero net moment. For the application of B causes the domains to reorganize themselves, and the direction of domain wall movement is reversed when B is reversed.

At the ferromagnetic transition the new permanent moment is accompanied by *magnetostriction* – a change in dimensions linked to the magnetization. As a result, iron which is cubic above T_c becomes slightly tetragonal, the cube axis (the direction of easy magnetization) along which M lies becoming elongated while the transverse cube axes contract. The effect is very small (about 1 part in 10^5) but this is irrelevant to the principle – a crystal either has cubic symmetry or it does not. No matter how close to unity the c/a axial ratio may be, if it is not precisely unity the crystal is not cubic. Landau recognized in developing his general theory that a λ-transition may be characterized by a qualitative and a quantitative change (note that *qualitative* does not mean inexact, as it is so frequently used by physicists). The qualitative change is the change of symmetry – it was cubic and now is tetragonal, and the change took place at a perfectly definite point, T_c. Simultaneously the order parameter, which was strictly zero above T_c, began to take a non-vanishing value, $\eta = c/a - 1$, say. The essence of the λ-point is that it marks the division between zero and non-zero values of η.

It must be stressed, however unnecessary such stress might seem, that there is no question of a cubic crystal, which may at some lower temperature become tetragonal, possessing in unobservable measure a germ of tetragonality before the change occurs: any more than an amoeba, which might have been the evolutionary forerunner of a vertebrate, can be said to possess a rudimentary backbone. And similarly, when η in Landau's theory takes on a non-vanishing value, this is not the enhancement of an unobservably small value that was present above T_c. Indeed, we have already seen in (6.4) how two branches come about, when the minimum condition for F demands that $\eta = 0$ or $(-\alpha/\beta)^{\frac{1}{2}}$. Above T_c, $\eta = 0$ is the only solution and is stable; below T_c this root describes an unstable solution and there is a switch to the other root, $\eta = (-\alpha/\beta)^{\frac{1}{2}}$. There is no mathematical continuity between solutions above and below T_c, and hence no justification in regarding the disordered state as one in which order happens to be very weak.

187

Although Landau's argument, as given here, was part of his over-simplified theory of the λ-point, its validity does not depend on this. Whatever the complexities arising from fluctuations there is still a precise temperature, T_c, below which long-range order is found; and it is the appearance of long-range order that is revealed by the symmetry of the whole crystal undergoing a sudden change. The fluctuating short-range order may lead to local strains but these will average out in a massive sample. One may also note that the switch of roots in (6.4) has no precise analogy in a true λ-transition, but the infinite peak to which C rises means that all the thermodynamic functions have a singular point at T_c, destroying any mathematical continuity between states above and below.

When a cubic crystal is transformed homogeneously into a tetragonal crystal, by one of the cube axes becoming different in length from the other two, there are three equivalent tetragonal forms possible according to which cube axis becomes the odd one out. The symmetry operation of rotation through $90°$ about one of the cube axes, that transforms a cube into itself, is retained by only one of the axes of the tetragonal crystal; but the ghost of the others is still there – rotation about either of the other two axes produces an equivalent tetragonal crystal which could just as well have been produced at the λ-point. This is a general property of λ-transitions: the higher symmetry on one side of the critical point remains on the other side as a potentiality, in the sense that all possible equivalent forms reveal the lost symmetry. The homogeneous crystal, however, after the transformation must select one only of the range of possibilities, and its symmetry is thereby lowered.

Exactly the same effect will be recognized in all the examples we have noted of stable or unstable symmetric transitions. Thus the perfect Euler strut may heel over either to the right or to the left after the critical load is exceeded; and thermal convection in a symmetrical circuit of pipework may choose either right- or left-handed flow. A slightly different example is provided by the instability of a long cylinder of liquid. Here the symmetry which is lost is the translational symmetry whereby bodily movement of the cylinder along its length leaves it unchanged (if end effects are negligible). The symmetry which is present before the instability operates shows up afterwards in the fact that the droplets may appear anywhere along the cylinder. It is worth while examining the other examples throughout this book to verify that the same principle applies to all. It can be summed up by the basic model of a stable symmetric transition in which a particle is subject to a constraint possessing left–right symmetry, $V = -ax^2 + cx^4$; if $a > 0$ the symmetry must be broken by either choice of stable solution, but applying the symmetry operation of reflection about the origin generates the other choice. Note that this is not a general property of the cusp catastrophe, where the potential need not have mirror symmetry, and the two alternative configurations need not be equivalent; only when $\varepsilon = 0$ is there a symmetry to be broken. A ferromagnet in zero external field is such a system, but if a field is applied the magnetic moment rises steadily as the temperature falls, and there is no point at which any discontinuity occurs. This is analogous to a cusp catastrophe when $\varepsilon \neq 0$.

14

122

154

156

Intrinsic broken symmetry

It is not necessary for a system to exhibit a critical point in order to show broken symmetry – the state of unbroken symmetry may be inaccessible. Consider, for example, the patterns taken up by soap films attached to various symmetrical arrangements of wires. If you drill holes in two transparent plastic sheets, and push wires through, you should be able to reproduce the patterns on the top row of fig. 7.1 (1-mm-diameter wires on a circle of 2 cm radius, between sheets 1 cm apart, work well). Only with three wires forming an equilateral triangle does the film possess the full symmetry. Otherwise the requirement that only three films meet along a line, and that they meet at 120°, demands that the symmetry be broken. With the square and the hexagon there are two equivalent solutions, with the pentagon there are five. On the bottom row of fig. 7.1 are two examples of three-dimensional film structures. The little square that forms in the middle of the wire cube may be parallel to any cube face, and one possibility is easily transformed into another by blowing gently on it. The beautiful pattern that can be made with a wire octahedron is only one of many, and none of them have octahedral symmetry. It is quite easy to make a small octahedron of wire, and I recommend you to make one, for the various patterns it supports are all pleasing to the eye, if nothing else.

The symmetry properties of a free atom, or of a gas of interacting atoms, are changed when they condense into a crystalline solid. There is nothing in the atom to indicate that argon will condense into a close-packed lattice with twelve nearest-neighbours to every atom. This is the consequence of the attractive forces between the atoms, which have no directional properties, and the fact that the close-packed hexagonal and the face-centred cubic lattices give the closest possible packing. But this latter fact is something quite separate from the nature of the force law, and nobody looking at the expression for the potential energy of the atoms (a function of their separations alone) could deduce from that alone what the crystal structure would be. It requires a separate geometrical investigation to find which crystal forms have the highest density, and therefore are most likely to be favoured by structureless atoms. This exemplifies a general feature of problems involving broken symmetry – if the symmetry properties of the

Fig. 7.1. Top row – soap films on parallel wires arranged in regular polygons; all films meet at 120°. Bottom row – soap films on a wire cube and octahedron.

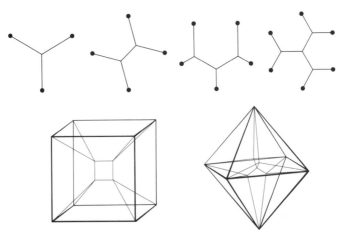

solution are not explicit in the statement of the problem there are no mathematical processes that will automatically generate the solution. Thus, even if the structureless force-law between argon atoms implies that the closest-packed crystal will be the stable form, the only way to discover which is most closely packed is to try all the possibilities. In this case the ultimate choice is limited to two, hexagonal and face-centred cubic, and which of these is the better is rather a subtle matter, and not our concern. One sees the problem most starkly when the atoms bond covalently, so that certain angles between bonds are favoured, and there is competition between close packing and the preferred configuration of neighbours. The problem of trying all likely possibilities is then really daunting, especially if one knows how very complicated the true answer may be; for instance, rhombohedral sulphur crystallizes with 128 atoms in its unit cell. The truth is that the *a priori* determination of crystal structures from the properties of the atom is almost always a fruitless endeavour; the best one can hope for is to find the structure by X-ray crystallography and to recognize how it reflects what is known about the preferred arrangement of small groups of atoms.

The same applies to chemical compounds. Only a very simple molecule, like methane, CH_4, can be said to reflect the properties of the carbon and

Fig. 7.2. Helical arrangement of silicon atoms in α-quartz (crystalline SiO_2). The oxygen atoms have been omitted, and the silicon atoms represented as rather large spheres so that the arrangement of the layers is clearer. Each pattern *abc* repeats to form a helix whose axis is shown as a dot. The lozenge of broken lines is the unit cell.

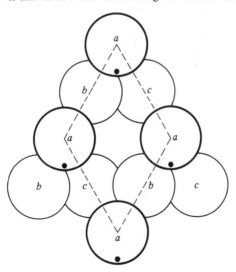

hydrogen atoms so closely that a theoretician could conceivably deduce its existence without empirical evidence. Everything else has so far lost the symmetry of the constituent atoms that it would be necessary (but hopeless) to solve Schrödinger's equation for the ground state energy of every imaginable compound, including its precise geometrical configuration, to discover by pure reason which stable forms could exist. This is not the way chemistry developed, and there is no merit in the claim made occasionally by physicists, that chemistry 'is only a branch of physics'. Once you realize that no one has yet deduced the structure of the water molecule, H_2O, from first principles, you appreciate how wild that claim is.

There is one form of broken symmetry exhibited by crystals and by chemical compounds that is closely parallel to the behaviour of the simple systems in chapter 5. This is the enantiomorphism which results in mirror-image pairs of compounds or crystals. For example, individual silica molecules, SiO_2, are identical with their mirror images, but in quartz they arrange themselves in parallel helices, all of the same handedness* (fig. 7.2). The corresponding mirror image has the same energy, and indeed both forms are commonly found coexisting in nature. Many organic molecules such as lactic acid, $CH(CH_3)(OH)COOH$, have four different groups attached to a carbon atom (sucrose is another). There are two equivalent ways, mirror images, of arranging the groups. For if we label them A, B, C, D and set them at the corners of a regular tetrahedron with a carbon atom in the middle, then looking from A towards the carbon atom we may see B, C, D running either clockwise or anticlockwise. All living creatures show handedness in their biochemistry – the sugar D-glucose is found throughout the animal kingdom, but its mirror image L-glucose is unknown except by laboratory synthesis. There are indeed few exceptions to the rule that anything, from lactic acid to the double helix of DNA, that has handedness will only occur biologically in one form. A mirror-image biology could presumably have developed just as well, and it is a nice point for speculation why only one is now extant.

---

* A right-handed screw thread is right-handed, whichever end you try to screw the nut on; seen in a mirror it looks like a left-handed screw thread. When plane polarized light is sent along a quartz crystal, parallel to the axes of these helices (*optic axis*), the plane of polarization is rotated, clockwise by one form and anticlockwise by the other. The rotation is in the same sense whichever direction the light travels. Optical activity like this is one way of recognizing enantiomorphism. When the enantiomorphism is a property of the individual molecule (e.g. sucrose), rotation of the plane of polarization occurs even with the molecule in solution.

Broken symmetry in quantum mechanics

So far I have used classical language to describe chemical compounds, portraying them with a sharply defined geometry. But whereas classical physics allows two equivalent structures, mirror images, for instance, to exist independently, quantum physics is not so accommodating. Let us return to the particle in a symmetrical potential, $V = -ax^2 + cx^4$ with $a > 0$ so that there are two equivalent solutions of minimum energy; we call this the symmetrical double-well model. The stationary wave functions obtained by solving Schrödinger's equation for this potential (which is a simple computational problem) have the property, common to all wave functions in a symmetrical potential, that they are either symmetric (s) or antisymmetric (a) with respect to the centre of symmetry of V (see fig. 7.3). We may indicate the character of the wave functions by the sign and magnitude of ψ in the two wells: $s = (1/\sqrt{2})(1, 1)$, $a = (1/\sqrt{2})(1, -1)$. The energy of s is normally lower than that of a, and we shall represent the energy difference by 2Δ. There is, of course, no suggestion that when the system is in a stationary state, either s or a, we may find the particle simultaneously in both wells; only that there is an equal probability of finding it in one well or the other if we look. And if we have found it in the left, our new knowledge allows us to write the wave function as $(1, 0)$, or $\frac{1}{2}(1, 1) + \frac{1}{2}(1, -1)$, both s and a states now being occupied equally. Because of the energy difference the phase difference between the two states changes at a rate $2\Delta/\hbar$, so that at an instant $\pi\hbar/\Delta$ later they will be in antiphase, $\frac{1}{2}(1, 1) - \frac{1}{2}(1, -1)$ or $(0, 1)$; the particle will have tunnelled through to the other minimum, only to go back and forth at a frequency $2\Delta/\hbar$ until the next observation disturbs the evolution of the wave function. As we allow a to pass from negative to positive, there is no critical behaviour when $a = 0$. As fig. 7.4 shows, the spacing of the levels quickly becomes very close, an indication of the rapidly increasing difficulty experienced by the particle in tunnelling through the central barrier. But at no point can we say that the description in terms of a split ground state suddenly ceases to be valid.

On the other hand, it is only pedantic to maintain the quantum point of view when the transition across the barrier is so slow as to be inappreciable, and when the ordinary processes of observing the state do not affect it. Thus a bar magnet magnetized in one sense has the same energy, in the absence of an applied magnetic field, as when its magnetization is reversed, and the quantum mechanical pedant could insist that it is really represented by the superposition of two states, the symmetrical ground state in which both senses are represented, separated from the antisymmetrical state by an energy considerably less than 10^{-51} J, so that the time for spontaneous reversal is longer than the age of the Universe. Between ridiculous pedantry and the essential invocation of quantum reasoning there is no mathematical frontier, but that does not mean that a very reasonable working compromise is ruled out.

To proceed towards a resolution of the apparent conflict it is convenient to take a specific case which illustrates how classical arguments can be

Fig. 7.3. The upper curve shows the potential $V = -ax^2 + cx^4$ and the lower curve the two lowest stationary solutions of Schrödinger's equation, symmetric (s) and asymmetric (a). By writing $a = (h^2 c^2/2m)^{1/3} A$, $E = (h^4 c/4m^2)^{1/3} \varepsilon$ and $x = (h^2/2mc)^{1/6} \xi$, the equation can be cast in the dimensionless form
$$\psi'' + (\varepsilon + A\xi^2 - \xi^4)\psi = 0.$$
For the case $A = 4$, as shown, $\varepsilon_s = -1.710\,457$ and $\varepsilon_a = -1.248\,098$.

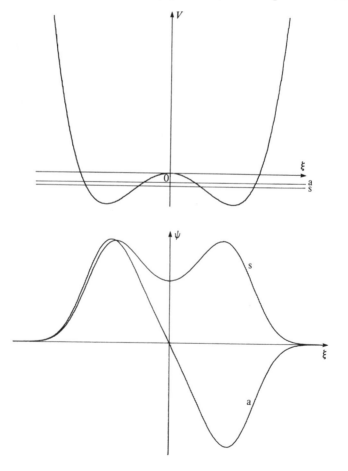

applied sometimes but not always. The ammonia molecule, NH_3, is customarily represented by a tetrahedron, formed of an equilateral triangle of hydrogen atoms with the nitrogen atom to one side of the hydrogen plane. If it were in the plane, at the centroid of the triangle, symmetry would demand that there be no permanent dipole moment. Yet the electrical susceptibility of ammonia gas varies inversely with temperature, just as a paramagnetic material whose molecules carry a permanent magnetic moment obeys Curie's law. The magnitude of the temperature-dependent susceptibility shows that the dipole moment of ammonia has the sort of size that one would expect from a tetrahedral arrangement – the nitrogen atom displacing some of the charge of the hydrogens in its direction. But of course if the tetrahedral arrangement is that of lowest energy, there is an equivalent mirror-image arrangement with the nitrogen on the other side of the hydrogen plane. The system is modelled by a charged particle moving in the double well, $V = -ax^2 + cx^4$, x being measured along the symmetry axis normal to the hydrogen plane. The existence of symmetrical and antisymmetrical states, each alone giving equal probabilities of finding the nitrogen on either side, is confirmed by the microwave absorption of ammonia at a frequency of about 24 GHz.[1] This shows that the levels are about 1.5×10^{-23} J apart, or alternatively that if the nitrogen is found at some instant to be on one side, it will be on the other side (and will not be found in its original position) after half a cycle, 4×10^{-11} s.

Fig. 7.4. Variation of ε_s and ε_a with A (see fig. 7.3 for notation). The broken lines show the asymptotic behaviour – when A is large and negative, $\varepsilon_s \sim |A|^{1/2}$ and $\varepsilon_a \sim 3|A|^{1/2}$; when A is large and positive, $\varepsilon_s = \varepsilon_a \sim -A^2/4 + (2A)^{1/2}$.

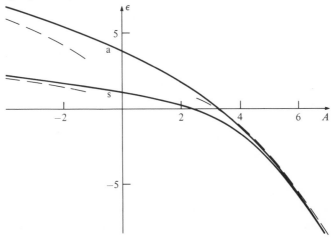

It is not obvious that the double-well model can simulate a permanent dipole moment, but in fact it can through the loss of symmetry that results from the application of an electric field. In a field $\mathscr{E}$ the particle, with charge q, is moving in the potential $-ax^2 + cx^4 - q\mathscr{E}x$, of the same form as (5.1) and describing an asymmetrical double well. The stationary states are perturbed so that the particle has not the same probability of being found in the two wells. The symmetrical state is changed to one possessing a dipole moment parallel to $\mathscr{E}$ and the antisymmetrical state to one possessing the same moment antiparallel.[2] If $\pm p$ is the dipole moment when the particle is in one well only, the dipole moments resulting from $\mathscr{E}$ are $\pm p^2\mathscr{E}/\Delta$; the lowest levels now have energy $\pm(\Delta^2 + p^2\mathscr{E}^2)^{\frac{1}{2}}$, instead of $\pm\Delta$.

Exercise 7.1. (a) Use the Boltzmann distribution to find the fraction of time spent in the two states, and hence the mean dipole moment when $\mathscr{E}$ is small, $\bar{p} = (p^2\mathscr{E}/\Delta)\tanh(\Delta/k_BT)$. This does not assume $\Delta \ll k_BT$, but if this is so, $\bar{p} \sim p^2\mathscr{E}/k_BT$.

(b) Compare this result with the behaviour of a permanent dipole p which can be oriented either parallel or antiparallel to $\mathscr{E}$, with energy $\pm p\mathscr{E}$. Show that the Boltzmann distribution leads to a mean dipole moment of $p^2\mathscr{E}/k_BT$ when $p\mathscr{E} \ll k_BT$.

This calculation shows that the classical permanent dipole and the polarizable double well in quantum mechanics show the same dielectric response so long as $k_BT \gg \Delta$. In ammonia the spacing $2\Delta = 2\pi\hbar \times (24\,\text{GHz}) = k_BT$, where $T = 1.2\,\text{K}$. The condition $\Delta \ll k_BT$ is very adequately met at room temperature, so that measurements of the susceptibility will not distinguish the classical and quantum models. There can be no objection to continuing to think classically, if it is easier, in these circumstances. It is, however, important to remember that the calculations of Ex. 7.1 assume complete thermal equilibrium, and that some very important applications depend on violating this condition. All masers and lasers operate by creating an inverted population in which the active system is more likely to be found occupying an excited state than some lower state, the reverse of the equilibrium Boltzmann distribution. And it is essential now to discuss the behaviour in terms of the real energy levels, and not any classical approximations. In the original ammonia maser,[3] for example, molecules in the upper, antisymmetric, state were separated from the rest by means of an electrostatic lens, and allowed to enter a cavity resonator, which they excited into electromagnetic oscillation by making downward transitions to the symmetrical state. Obviously the maser could not have worked unless the lens effectively inverted the population, and the molecules entered the cavity before radiation from the surroundings, or collisions, could restore the Boltzmann distribution. Again, the correct energy levels and wave functions must obviously be employed to understand the spectroscopic behaviour of ammonia and analogous systems. There is usually little difficulty deciding when to abandon the classical approximation.

Reconciliation of quantum and classical physics

It is all very well to say that you can easily decide when the classical approximation is safe, but this does not answer the pedant we dismissed so summarily a page or two back. The worlds of quantum and of classical physics are very different, and one may well agree with him in feeling uneasy about allowing intuition to determine which point of view to adopt. If it is true that classical mechanics is a limiting form of quantum mechanics, surely we ought to be able to analyse an obviously classical problem in quantum-mechanical terms, and exhibit explicitly the criterion for forgetting about quantum effects. When we embark on this task, however, we may not find it as straightforward as we were tempted to assume. Let us consider a specific case: I take a bar magnet, with differently coloured ends, from my pocket and show it to you, asking you to describe it in quantum-mechanical language. However you set about the task your wave function must be such as to allow both ends the same chance of being North. As soon as you hold a compass needle anywhere near you will know the answer, and your original wave function will immediately become invalid, needing to be replaced by one that describes the red end (say) as indisputably North. This is the process technically referred to as the reduction of the wave packet. So far, so good – but what if I have already made my own test and know privately that the red end is North? Was your original wave function wrong, because you failed to take account of what you couldn't have known – that I had already reduced the wave packet? It seems a strange way of solving problems if you have to start with information of whose existence you were ignorant! On the other hand, if you feel you must ignore such hypothetical information and hence claim that your original wave function was correct in principle, it gives you half a chance of finding, contrary to what I know, that the red end is South – an unlikely story.

We have got ourselves into a fix, and very much the sort of fix that has worried physicists, even as distinguished as Einstein and Schrödinger, ever since quantum mechanics was discovered.[4] There is, as far as anybody knows, only one way out, and that is to do the quantum mechanics properly, without any of the plausible short cuts we took in the last paragraph and which led us into a logical ambush. This will not explain anything – the goings-on at the quantum level will remain just as mysterious – but it will show that the paradoxes we have uncovered are not real, but only artefacts resulting from a misuse of theory. It was quite reasonable, but also quite wrong, to assume that you and I were independent observers. If we are ever to compare our observations we must include in the initial wave function the magnet and ourselves (or perhaps it will be enough to let the compass needles act for us). We shall then find that the quantum mechanics of this total system precludes us from getting different answers. We may both find the red end is North, or both find it South, but never can one find one and the other the other. The quantum world makes sense when applied to classical problems.

To demonstrate this conclusion I shall abandon the story of the magnet in favour of a simpler hypothetical experiment. Instead of a bar magnet I take a dipole consisting of a charged particle that may sit in either of two deep potential wells, with a fixed opposite charge between them. Thus the dipole may take either sign, according to which well the particle occupies, but there is a negligible chance of its tunnelling from one to the other. The lowest two states, s and a, are then so close in energy that we may treat them as degenerate, and take any linear combination as an equally good eigenfunction. For our purposes it is convenient to replace $(1/\sqrt{2})(1, 1)$ and $(1/\sqrt{2})(1, -1)$ by $(1, 0)$, which we call $L(x)$, and $(0, 1)$, which we call $R(x)$, x being the position coordinate of the particle. If we have prepared the system so that it may equally well be in either state, we can represent its wave function as $(1/\sqrt{2})(L+R)$.* At this stage we have not allowed ourselves any observation of which well is occupied, but now we do indeed carry out a test.

To prepare a suitable test object we use an electrostatic lens to isolate molecules of ammonia in their excited state, $\chi_a = (1/\sqrt{2})(1, -1)$. We shoot them close to a dipole, similar to that under investigation, but one whose polarity we know, and let them pass by the unknown dipole immediately after. By adjusting the closeness of interaction, and its duration, it is

---

* This is a very rough-and-ready expression for the wave function. Strictly we should use density matrices but as the result will be the same the extra complication may be dispensed with. Perhaps we ought to add an unknown phase difference, writing $(1/\sqrt{2})(L + Re^{i\varphi})$ instead of $(1/\sqrt{2})(L+R)$, but again it does not change the result and the simplest schematic wave function suffices to reveal the argument.

possible to arrange that the wave function of a molecule suffers a $\pi/2$ phase shift, i.e. $(1/\sqrt{2})(1, -1)$ becomes $(1/\sqrt{2})(1, i)$ on passing a dipole in state R, and $(1/\sqrt{2})(1, -i)$ if the dipole is in state L. Then if the two dipoles are in the same state, the molecule is switched from a to s, $\chi_s = (1/\sqrt{2})(1, 1)$, while if they are in opposite states it remains in a. Since an electrostatic lens can sort a and s molecules into different bins, we need only look at which bin the test molecule arrives at to know whether the two dipoles were the same or opposite. So long as the tunnelling barrier in the test molecules is much lower than that of the dipole, the test procedure runs no risk of altering the polarity of the dipole.

Expressed formally, the initial wave function of the combined system, dipole + test molecule, is $(1/\sqrt{2})(L + R)\chi_a(y_1)$, where y_1 is the coordinate representing the configuration of the test molecule. The test procedure turns $L\chi_a$ into $L\chi_s$ if the known dipole is in state L, while $R\chi_a$ remains $R\chi_a$. Thus after the test passage the wave function is $(1/\sqrt{2})(L\chi_s + R\chi_a)$, and the fraction represented by $(1/\sqrt{2})L\chi_s$ finds its way to the s-bin, while the fraction $(1/\sqrt{2})R\chi_a$ finds its way to the a-bin. The intensity in each bin is $\frac{1}{2}$, indicating that we have equal chance of finding that the dipole was L or R.

Now let a second test molecule pass through. In accordance with the rule that every contributing system must be included in the wave function, we write the initial state as $(1/\sqrt{2})(L + R)\chi_a(y_1)\chi_a(y_2)$. The passage of the first molecule turns this into $(1/\sqrt{2})(L\chi_s(y_1)\chi_a(y_2) + R\chi_a(y_1)\chi_a(y_2))$, and the passage of the second molecule completes the transformation into $(1/\sqrt{2})(L\chi_s(y_1)\chi_s(y_2) + R\chi_a(y_1)\chi_a(y_2))$. The first term tells us that there is half a chance of both molecules going to the s-bin, and the second term that there is half a chance of both going to the a-bin. In the absence of cross-terms like $\chi_s(y_1)\chi_a(y_2)$ there is no possibility that one molecule will end in each bin.* The process can be repeated indefinitely – however many tests are made, they will all give the same answer, either L or R, but never a hint of ambiguity. The quantum mechanical analysis has shown the dipole to possess one essential property of classical systems, that all observers will

---

* Notice that this result only comes about when both observers (i.e. test molecules) are incorporated in the single wave function. If the end-point for the dipole and first observer could be described by $(1/\sqrt{2})(L\chi_s(y_1) + R\chi_a(y_1))$ and the dipole and second observer by $(1/\sqrt{2})(L\chi_s(y_2) + R\chi_a(y_2))$, each would find an equal chance for either outcome, but they would be uncorrelated – half the time they would agree and for the rest they would disagree. It was the intrinsic unlikeliness that the two observers, however far apart in time and space their observations, would have to be treated as part of the same system if they were ever to compare results, that led (among other difficulties) to Einstein's antagonism to the accepted Copenhagen interpretation.

agree on its configuration. For this result to hold, here and elsewhere, it is necessary that the system may be examined without changing its configuration, and that the configuration itself shall be stable, suffering no spontaneous change during the course of the test.

But is there anything more to what we have done than the trivial interpretation to the effect that the dipole *really* points in a definite direction, and all this paraphernalia of quantum mechanics is only a cloak for our ignorance of which direction it is? Yes, there is more than that – what looks like a sensible man-in-the-street attitude to quantum mechanics simply won't work. The idea that electrons, etc., are pursuing a perfectly definite path, which the grossness of our senses prevents us from seeing, is an example of a *hidden-variables* interpretation, such as von Neumann showed was mathematically incompatible with the form taken by quantum mechanics. Whatever is going on, if indeed anything is going on, in the submicroscopic world, it is not such as can be pictured in terms of the experiences of the everyday world. The procedure I have gone through, solving the quantum problem without short cuts, demonstrates how, out of this inconceivable substructure, there is projected in suitable circumstances a form of behaviour which accords with normal experience and which I have termed classical – the physics of things which are big enough to submit without significant change to the scrutiny of many observers, and which we now find will present the same aspect to all. One can go further, and show that successive observations of a dynamical system, when analysed in similar fashion with all observers included in the same wave function, will lead to the appearance of a definite trajectory, obeying Newton's laws of motion.[5] But this is outside our present scope; it is enough to have shown how the broken symmetry of classical systems and the apparently unbroken symmetry of the quantum treatment of the same systems are compatible; and it is the basic property of classical systems – that of presenting the same appearance to all observers – that resolves the differences of description.[8]

Limitations of science[6]

[I observed, a few pages back, that the phenomenon of broken symmetry disposed of the facile view that chemistry was a branch of physics. This is no disparagement of physics – it is a great triumph to have elucidated the laws governing fundamental particles, and to understand the origin of the valence forces that bond atoms in materials. One can confidently claim that nothing in chemistry casts doubt on the fundamental principles of physics, or demands the introduction of new laws.

But, although the laws of physics suffice to explain, they give little guidance towards construction; they are too general for that, implying as they do infinitely more structures that *might be* than the number that actually *is*. All the fascinating forms we observe result from limitation of the range of possibilities, when the full symmetry of the equations is replaced by something of lower symmetry. The only systematic procedure we know to tackle the diversity of forms is to observe and classify; after that the reductionist process may begin, the search for underlying rules and, ultimately, the discovery of the fundamental laws from which all else springs. It is fashionable in anti-intellectual circles to sneer at reductionism, as if there were an alternative approach to understanding. But those who claim to take a 'holistic' view are at best natural historians who try to observe more closely and classify more precisely; at worst they are charlatans imposing pretentious fictions on the gullible. Of course every physicist wishes it were possible to construct *ab initio* the real world from the fundamental laws, but few have the imagination to make even a tentative start, and without imagination nothing is possible, since systematic procedures are unavailable.

The history of science reflects this technical limitation. It began as a number of separate studies, impelled by practical needs and with no thought of an all-embracing comprehension of the material world. Medicine, herb-lore and husbandry after thousands of years developed into biology; metal-working, weapons and the lust for gold into chemistry and metallurgy. Progress followed no overall pattern but occurred where progress was possible, and there was little or no correspondence between the different arts. The intellectual landscape is like that of Papua, fertile valleys separated by well-nigh impassable mountains and forests, so that each tribe has developed its own language and customs in isolation. Even now it is only rarely that the tribes master their own territories to the point of driving a pathway to their neighbours, only to find that communication is far from easy. Nevertheless, the physicists are beginning to appreciate in their own way the peculiar methods of the chemists, and the chemists beginning to see the relevance of their own knowledge to the study of living organisms.

It is not merely between different disciplines that contact is hard to achieve; within a single discipline there are critical discontinuities which enforce different languages and outlooks on those working on either side. The solid-state physicist constructs heuristically, rather than deriving, his model of a solid. In a rather different way the biological world exhibits untold numbers of critical points, as the evolution of complex organisms brings about the appearance of properties previously not even hinted at. Thus the simplest cellular creatures are apparently innocent of sex, while under the right conditions some bacteria can multiply sexually rather than by cell division. These latter possess the capability of sex, however infrequently they may use it – more primitive cells have no capability at all this way. Just as in a λ-transition, the critical point is when the capability appears, however rarely it may be exercised. And there must be very many properties of which the same may be said. In particular, at a rather late (very late, perhaps) evolutionary stage a new form of behaviour arises – consciousness, or what may be a special kind, self-awareness. This is a characteristic of certain very highly organized creatures, but the extreme difficulty, if not impossibility, of studying it objectively in a scientific manner leaves one baffled to determine what are the conditions for its occurrence. Courtesy bids us assume that most of mankind are like us, but beyond that we are in doubt. A stone is not aware of itself, or a virus or a worm, but what about a cat, a chimpanzee or a dolphin? Fortunately I am not called upon to decide in any specific case – it is enough to recognize that the world of beings can be divided into those that have no intrinsic capacity to know themselves, and those with that capacity, however feeble. And presumably at some stage in the evolution of Man there was a critical mutation conferring this new, and to the physicist wholly mysterious, capability.

The word *mysterious* deserves to be stressed. With phase changes and other critical transitions in the physical world we may be unable to predict what happens on the other side, but when we observe it we recognize that it conforms to the physical laws and can be understood to the same degree. The property of awareness is something different; if science is concerned with those matters of observation which are common to all reasonable observers, awareness is not a proper object of scientific study – only I can observe my own awareness, and my description of it is subjective and not susceptible to independent confirmation. This does not mean that the processes in the brain are not in accordance with the laws of physics – personally I believe that any objective study of the processes accompanying thought will show them to be similar to other biological processes, though perhaps more complicated; but no such study can take us out of the objective into the subjective. Here scientific rationalism meets its match.

There is nothing new in this conclusion;[7] the 'mind versus matter' conundrum has been with us since the Ancient Greeks, and is no nearer resolution now than then. Physics will give no help except possibly, by clarifying the nature of critical phenomena, to sharpen the debate; and it is desirable that scientists should recognize their limitation. If we cannot understand the profound change that takes place at the critical point where self-awareness enters, we are in no position to make dogmatic assertions about what other new phenomena may accompany it. In particular, the religious experience is not to be dismissed as incompatible with what we know of the material world, when the very fact that we know we can experience the material world is itself incompatible with what we know of that world. Nevertheless the scientist has a part to play in religious disputation because all great religions carry a fair amount of accessory baggage in the form of cosmological theories. It is proper for a scientist to intrude his own cosmology (I use the term very generally), as for example by parading the evidence for evolution as an alternative to special creation. But there is no scientific warrant for questioning anyone's belief on the relationship between himself and his god. One wishes these two assertions could be generally agreed.]

REFERENCES

There are a number of references to A. B. Pippard, *The Physics of Vibration*, vol. 1 (1978: Cambridge University Press). This book will be denoted below by *PV.*

CHAPTER 2

1 The theory of the Euler strut will be found in almost every book of the sort that used to be called 'Properties of Matter', e.g. the book of that name by F. C. Champion and N. Davy, p. 56 (1937: Blackie). Also books on engineering structures and strength of materials, e.g. A. J. S. Pippard and J. F. Baker, *The Analysis of Engineering Structures*, 3rd edn, p. 114 (1957: Arnold).
2 Euler's equations for the motion of a rigid body are proved in almost every book on Rigid Body Dynamics, e.g. H. C. Corben and P. Steble, *Classical Mechanics*, 2nd edn, p. 49 (1960: Wiley); or A. B. Pippard, *Forces and Particles*, p. 119 (1972: Macmillan).
3 For a proof, see B. M. Brown, *The Mathematical Theory of Linear Systems*, p. 244 (1961: Chapman & Hall).
4 For a proof, see H. W. Bode, *Network Analysis and Feedback Amplifier Design*, p. 137 (1945: Van Nostrand).
5 e.g. D. K. Anand, *Introduction to Control Systems*, 2nd edn (1984: Pergamon).

CHAPTER 3

1 e.g. E. R. Andrew, *Nuclear Magnetic Resonance* (1956: Cambridge University Press).
2 For a sample of derivations and discussions of the Kramers–Kronig relations see H. Fröhlich, *Theory of Dielectrics*, p. 2 (1949: Clarendon Press); L. D. Landau and E. M. Lifshitz, *Statistical Physics*, p. 391 (1958: Pergamon); or *PV*, p. 109.
3 *PV*, p. 142.
4 See any book on wave optics, e.g. S. G. Lipson and H. Lipson, *Optical Physics*, p. 152 (1969: Cambridge University Press).
5 K. F. Riley, *Mathematical Methods for the Physical Sciences*, p. 182 (1974: Cambridge University Press).
6 Ref. 3.4, p. 175.
7 *PV*, p. 44.
8 Ref. 3.5, p. 205.
9 C. Kittel, *Introduction to Solid State Physics*, 4th edn, p. 239 (1971: Wiley). .
10 Ref. 3.5, p. 214.
11 P. Debye, *Polar Molecules*, p. 77 (English version 1945: Dover); see also H. Fröhlich, ref. 3.2, p. 83.

12 P. Vigoureux, *Ultrasonics*, p. 78 (1950: Chapman & Hall).

13 T.-S. Kê, *Phys. Rev.* **74**, 9 (1948).

14 C. F. J. Böttcher and P. Bordewijk, *Theory of Electric Polarization*, vol. 2, fig. 49 (1978: Elsevier).

15 K. S. Cole and R. H. Cole, *J. Chem. Phys.* **9**, 341 (1941).

16 R. W. Leonard, *J. Acoust. Soc. Am.* **12**, 241 (1940).

17 A. A. Maryott, A. Estin and G. Birnbaum, *J. Chem. Phys.* **32**, 1501 (1960).

18 See e.g. R. L. Sproull and W. A. Phillips, *Modern Physics*, 3rd edn, p. 258 (1980: Wiley).

19 A. Einstein, *Sitz-Ber. Preuss Akad. d. Wiss*, p. 380 (1920).

20 J. W. Strutt, *Proc. London Math. Soc.* **4**, 357 (1873) (*Collected Works of Lord Rayleigh*, vol. 1, p. 170); J. R. Carson, *Proc. I.R.E.* **17**, 952 (1929).

21 A. J. S. Pippard and J. F. Baker, ref. 2.1, p. 53.

22 A. B. Pippard, *The Elements of Classical Thermodynamics*, p. 45 (1957: Cambridge University Press).

23 S. R. de Groot, *Thermodynamics of Irreversible Processes*, p. 5 (1951: North-Holland).

CHAPTER 4

1 *PV*, p. 323.

2 J. K. Roberts and A. R. Miller, *Heat and Thermodynamics*, 5th edn, p. 105 (1960: Blackie). But more recent authors seem to avoid this calculation or do it otherwise, e.g. A. J. Walton, *Three Phases of Matter*, p. 119 (1976: McGraw-Hill).

3 Results obtained with a non-linear electrical circuit are beautifully illustrated in H. J. Janssen, L. Beerden and E. L. M. Flerackers, *Eur. J. Phys.* **5**, 94 (1984).

4 J. J. Stoker, *Nonlinear Vibrations*, chap. 4 (1950: Interscience), is fairly straightforward. A. H. Nayfeh and D. T. Mook, *Nonlinear Oscillations*, chap. 4 (1979: Wiley), is more up-to-date but over-detailed as far as most readers are concerned.

5 *PV*, p. 385.

6 *PV*, p. 388.

7 A. H. Benade, *Fundamentals of Musical Acoustics*, p. 387 (1976: Oxford University Press).

8 A. B. Bronwell and R. E. Beam, *Theory and Applications of Microwaves*, chap. 8 (1947: McGraw-Hill).

9 B. Lax and K. J. Button, *Microwave Ferrites and Ferromagnetics*, p. 563 (1962: McGraw-Hill).

10 D. R. Hamilton, J. K. Knipp and J. B. H. Kuper, *Klystrons and Microwave Triodes*, p. 414 (1948: McGraw-Hill).

11 Part of a letter from C. Huygens to his father is quoted in M. Sargent, M. O. Scully and W. E. Lamb, *Laser Physics*, p. 52 (1974: Addison-Wesley).

12 *PV*, p. 391.

13 The same equation describes a superconducting Josephson junction; see T. Van Duzen and C. W. Turner, *Principles of Superconductive Devices and Circuits*, chap. 4 and p. 170 (1981: Arnold).

14   J. J. Stoker, ref. 4.4.

15   *PV*, pp. 253 and 271.

16   For an elementary account of matrix algebra, see ref. 3.5, chap. 14.

17   R. M. May, *Nature*, **261**, 459 (1976). This short paper was most physicists' introduction to the complexities of one-dimensional maps, and must have been an important stimulus to the development of the subject outside pure mathematics.

18   P. Cvitanović has edited and written an introduction to a collection of fundamental papers in this field – *Universality in Chaos* (1984: Hilger). Two papers by M. J. Feigenbaum will be found on pp. 49 and 207.

19   W. Shakespeare, *Othello*, Act 3, scene 3, l. 92.

20   Ref. 4.18, p. 30. See also J. E. Hirsch, B. A. Huberman and D. J. Scalopino, *Phys. Rev. A*, **25**, 519 (1982).

21   *PV*, p. 403.

22   Ref. 3.5, chap. 11.

23   Ref. 4.18, p. 341. M. Hénon's model obeys the equations
$$X_{n+1} = Y_n + 1 - aX_n^2 \quad \text{and} \quad Y_{n+1} = bX_n.$$
This can be computed very rapidly; try for yourself with the values used by Henon, $a = 1.4$, $b = 0.3$.

24   B. B. Mandelbrot, *The Fractal Geometry of Nature*, chap. 20 (1977: Freeman).

25   B. A. Huberman and J. P. Crutchfield, *Phys. Rev. Lett.* **43**, 1743 (1979).

26   R. L. Kautz, *J. App. Phys.* **52**, 6241 (1981).

27   R. W. Rollins and E. R. Hunt, *Phys. Rev. Lett.* **49**, 1295 (1982), give references to earlier descriptions of the circuit in fig. 4.28(*a*) and develop a theory. The circuit of fig. 4.28(*b*) is described by E. R. Hunt, *Phys. Rev. Lett.* **49**, 1054 (1982). A quite different, and easily assembled, chaotic circuit is described by J. P. Gollub, T. O. Brunner and B. G. Danly, *Science*, **200**, 48 (1978).

28   H. Bénard, *Ann. d. Chim. et d. Phys.* **23**, 62 (1901).

29   Lord Rayleigh, *Phil. Mag.* **32**, 529 (1916).

30   G. Ahlers, *Phys. Rev. Lett.* **33**, 1185 (1974); D. C. Threlfall, *J. Fluid Mech.* **67**, 17 (1975).

31   Ref. 4.18, p. 114. A good general review of Rayleigh–Bénard convection is given by P. Berge and M. Dubois, *Contemp. Phys.* **25**, 535 (1984). See also P. Drazin and W. Reid, *Hydrodynamic Stability*, p. 51 (1981: Cambridge University Press).

32   Ref. 4.18, p. 109.

33   I am sure I have seen this remarked on somewhere, but cannot remember where.

34   I. and R. Opie, *The Oxford Dictionary of Nursery Rhymes*, p. 324 (1952: Clarendon Press).

CHAPTER 5

1   R. V. Southwell, *An Introduction to the Theory of Elasticity*, p. 471 (1936: Clarendon Press).

2   K. S. Krishnan and S. Banerji, *Phil. Trans. Roy Soc. A*, **234**, 267 (1935).

3   Other multivibrators are discussed in some detail in *PV*, p. 353.

4   For a more formal treatment, L. A. Pars, *A Treatise on Analytical Dynamics*, p. 214 (1965: Heinemann).

5   *PV*, chap. 10.

6   A. B. Pippard, *Eur. J. Phys.* **1**, 13 (1980).

7   Ref. 3.5, p. 356.

8   A. C. Palmer and J. H. Martin, *Nature*, **254**, 46 (1975).

9   J. W. Strutt, *Phil. Mag.* **14**, 184 (1882). (*Collected Works of Lord Rayleigh*, vol. 2, p. 130.)

10   E. M. Hartley, *Cartesian Geometry of the Plane*, p. 174 (1960: Cambridge University Press).

11   J. H. Jeans, *The Mathematical Theory of Electricity and Magnetism*, 5th edn, p. 248 (1946: Cambridge University Press).

12   J. Zeleny, *Phys. Rev.* **10**, 1 (1917).

13   G. I. Taylor, *Proc. Roy. Soc. A*, **280**, 383 (1964).

14   H. N. V. Temperley, *Properties of Matter*, p. 94 (1953: University Tutorial Press).

15   M. Van Dyke, *An Album of Fluid Motion*, p. 73 (1982: Parabolic Press). This very beautiful book is worth consulting for the many other examples of fluid instability it illustrates.

16   J. W. Strutt, *Proc. Lond. Math. Soc.* **10**, 4 (1879). (*Collected Works of Lord Rayleigh*, vol. 1, p. 361.)

17   Books on catastrophe theory: T. Poston and I. Stewart, *Catastrophe Theory and its Applications* (1978: Pitman); P. T. Saunders, *An Introduction to Catastrophe Theory* (1980: Cambridge University Press); R. Gilmore, *Catastrophe Theory for Scientists and Engineers* (1981: Wiley). Saunders' book is the shortest and is noteworthy for its modest appraisal of the value of catastrophe theory. For a wide variety of applications see E. C. Zeeman, *Catastrophe Theory: Selected Papers 1972–1977* (1977: Addison-Wesley).

18   C. F. G. Delaney, *Electronics for the Physicist*, p. 145 (1969: Penguin).

19   K. Foster and G. A. Parker, *Fluidics*, p. 5 (1970: Wiley). If you want to know more about how the device works, see chap. 3 of this book or J. M. Kirscher, *Fluid Amplifiers*, chap. 12 (1966: McGraw-Hill).

20   Ref. 5.14, p. 145.

21   Ref. 4.2.

22   J. C. Maxwell, *Nature*, **11**, 357 (1875).

23   A. B. Pippard, *Elements of Classical Thermodynamics*, p. 108 (1957: Cambridge University Press).

24   R. B. Griffiths, *Phys. Rev.* **158**, 176 (1967), argues that Maxwell's construction is demanded by the analytical character of van der Waals' equation outside the two-phase region.

25   R. R. Rogers, *A Short Course in Cloud Physics*, 2nd edn, p. 55 (1979: Pergamon).

26   A. R. Ubbelohde, *The Molten State of Matter*, p. 404 (1978: Wiley).

27   T. Poston and A. E. R. Woodcock, *Proc. Camb. Phil. Soc.* **74**, 217 (1973). See also M. A. B. Whittaker, *Eur. J. Phys.* **4**, 212 (1983), and many more examples in ref. 5.17 and in J. M. T. Thompson, *Instabilities and Catastrophes in Science and Engineering* (1982: Wiley).

28   Ref. 5.17 and R. Thom, *Structural Stability and Morphogenesis* (1975: Benjamin).

29   J. F. Nye, *Proc. Roy. Soc. A*, **361**, 21 (1978).

CHAPTER 6

1 L. P. Kadanoff, W. Götze, D. Hamblen, R. Hecht, E. A. S. Lewis, V. V. Palcianskas, M. Rayl, J. Swift, D. Aspres and J. Kane, *Rev. Mod. Phys.* **39**, 395 (1967). This is a comprehensive review of phase transition data and theory, from which many of the experimental results quoted in this chapter are taken.

2 H. L. Lorentzen, quoted in ref. 6.1.

3 A. B. Pippard, ref. 2.2, p. 142.

4 S.-K. Ma, *Modern Theory of Critical Phenomena*, chaps. 4 and 5 (1976: Benjamin).

5 M. I. Bagatskii, A. V. Voronel' and B. G. Gusak, *Soviet Physics JETP*, **16**, 517 (1963).

6 T. Andrews, *Phil. Trans. Roy. Soc.* **159**, 575 (1869).

7 M. v. Smoluchowski, *Ann. d. Physik*, **25**, 205 (1908).

8 B. I. Bleaney and B. Bleaney, *Electricity and Magnetism*, 3rd edn, p. 239 (1976: Oxford University Press).

9 See L. Rosenfeld, *Theory of Electrons*, p. 80 (1951: North-Holland).

10 A. Andant, *Ann. de Physique*, **1**, 346 (1924).

11 M. J. Buckingham and W. M. Fairbank, *Prog. in Low Temp. Phys.* **3**, 80 (1961: North-Holland).

12 F. L. Lederman, M. B. Salamon and L. W. Shacklette, *Phys. Rev. B*, **9**, 2981 (1974).

13 H. Moser, *Phys. Z.* **37**, 737 (1936).

14 V. C. Reinsborough and F. E. W. Whitmore, *Aust. J. Chem.* **20**, 1 (1967).

15 S. A. Friedberg and D. Wasscher, *Physica*, **19**, 1072 (1953).

16 Drawn from data in W. S. Corak and C. B. Satterthwaite, *Phys. Rev.* **102**, 662 (1956).

17 L. Onsager, *Phys. Rev.* **65**, 117 (1944).

18 N. G. Parsonage and L. A. K. Staveley, *Disorder in Crystals*, p. 24 (1978: Oxford University Press).

19 R. W. Jones, G. S. Knapp and C. W. Chu, *18th Conf. on Magnetism*, 1618 (1973).

20 See L. D. Landau and E. M. Lifshitz, *Statistical Physics*, chap. 14 (1958: Pergamon).

21 W. L. Bragg and E. J. Williams, *Proc. Roy. Soc. A*, **145**, 699 (1934).

22 P. Weiss, *J. de Physique*, **6**, 661 (1907).

23 C. Kittel, *Introduction to Solid State Physics*, 4th edn, chap. 14 (1971: Wiley).

24 See N. W. Ashcroft and N. D. Mermin, *Solid State Physics*, p. 679 (1976: Holt, Rinehart and Winston).

25 Ref. 6.4, p. 67.

26 C. Kittel, *Elementary Statistical Physics*, p. 52 (1958: Wiley).

27 H. E. Hall, *Solid State Physics*, p. 71 (1974: Wiley).

28 K. H. Stewart, *Ferromagnetic Domains* (1954: Cambridge University Press).

29 F. Bitter, *Phys. Rev.* **38**, 1303 (1931).

30 D. Shoenberg, *Superconductivity*, chap. 4 (1952: Cambridge University Press).

31 C. E. Chase, *Phys. of Fluids*, **1**, 193 (1958).

32 I. Hatta, *J. Phys. Soc. Japan*, **28**, 1266 (1970).

CHAPTER 7

1 A. B. Pippard, *The Physics of Vibration*, vol. 2, p. 122 (1983: Cambridge University Press).

2 Ref. 7.1, p. 96.

3 J. P. Gordon, H. Z. Ziegler and C. H. Townes, *Phys. Rev.* **95**, 282 (1954).

4 J. C. Polkinghorne, *The Quantum World*, chap. 6 (1984: Longman).

5 N. F. Mott, *Proc. Roy. Soc. Lond.* **126A**, 79 (1929). This and many other basic papers on the interpretation of quantum mechanics will be found in J. A. Wheeler and W. H. Zurek, *Quantum Theory and Measurement* (1983: Princeton University Press).

6 This heading is the title of a book by J. W. N. Sullivan (1933: Chatto and Windus), a stimulating essay by the most distinguished scientific populariser of his time, which after fifty years still makes very good reading. Some of the topics touched on in this section will be found there also, especially on p. 159 and after.

7 C. Lloyd Morgan, *Emergent Evolution*, 2nd edn (1927: Williams and Norgate), has a lengthy discussion of related problems, with references to earlier philosophers.

8 A. J. Leggett, *Contemp. Phys.* **25**, 583 (1984) discusses related problems in some detail.

INDEX